D1222836

THE DEVELOPMENT OF NEWTONIAN CALCULUS IN BRITAIN 1700–1800

NICCOLÒ GUICCIARDINI

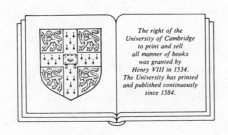

The right of the
University of Cambridge
to print and sell
all manner of books
was granted by
Henry VIII in 1534.
The University has printed
and published continuously
since 1584.

CAMBRIDGE UNIVERSITY PRESS

Cambridge

New York   Port Chester   Melbourne   Sydney

Published by the Press Syndicate of the University of Cambridge
The Pitt Building, Trumpington Street, Cambridge CB2 1RP
40 West 20th Street, New York, NY 10011, USA
10 Stamford Road, Oakleigh, Melbourne 3166, Australia

First published 1989

Printed in Great Britain by the University Press, Cambridge

British Library cataloguing in publication data
Guicciardini, Niccolò
The development of Newtonian calculus in Britain, 1700–1800
1.   Calculus, history
I.   Title
515.09

Library of Congress cataloguing in publication data

Guicciardini, Niccolò.
The development of Newtonian calculus in Britain 1700–1800/
Niccolò Guicciardini.
p. cm.
Originally presented as the author's thesis (Ph. D.–Council for
National Academic Awards, 1987)
Bibliography: p.
Includes index.
ISBN 0 521 36466 3
1. Calculus–Great Britain–History–18th century. I. Title.
QA303.G94 1989
515'.0941'09033–dc20 89-7085 CIP

ISBN 0 521 36466 3

CONTENTS

INTRODUCTION

EIGHTEENTH-CENTURY British mathematics does not enjoy a good reputation. The eighteenth century, a 'period of indecision'[1] as many historians would say, is said to have witnessed 'the crisis' or the 'decline' of mathematics in the country of Newton, Wallis and Barrow. However, even a glance at the following list of names should be sufficient to refute the prevailing image of eighteenth-century British mathematics. To the imported Abraham de Moivre one can add the native Brook Taylor, James Stirling, Edmond Halley, Roger Cotes, Thomas Bayes, Colin Maclaurin, Thomas Simpson, Matthew Stewart, John Landen and Edward Waring. Through their work they contributed to several branches of mathematics: algebra, pure geometry, physical astronomy, pure and applied calculus and probability.

I devote this work to a theory that all these natural philosophers knew very well: the calculus of fluxions. This was the British equivalent of the more famous continental differential and integral calculus. It is usually agreed that the calculus of fluxions was clumsy in notation and awkward in methodology: the preference given to Newton's dots and to geometrical methods engendered a period which was eventually labelled as the 'Dot-Age'.[2] Furthermore, the calculus of fluxions is usually indicated as the principal cause of the decadence of British mathematics: the 'Dot-Age' was the price paid for a chauvinistic attachment to Newton's theory.

The origin of this depressing image of the Newtonian calculus can be easily traced back to the irreverent writings of the Cambridge Analytical Society's fellows who, at the beginning of the nineteenth century, tried to introduce into Great Britain the algebraical methods of Lagrange and Arbogast.[3] Like all the reformers, they offered a pessimistic view of the past. Since then, many historians have behaved as loyal members of the Analytical Society, and a standard account of the eighteenth-century fluxional calculus has been given in the histories of mathematics. For

vii

instance, in Koppelman (1971) we find stated that the 'quiescence' of English mathematics in the eighteenth century depended upon the isolation of English mathematicians from the continent. The reason for this isolation is attributed to the 'bitterness engendered by the Newton–Leibniz priority controversy' and to the 'insularity of the English'. The result of this situation was, according to Koppelman, that 'the Newtonian school clung to a clumsy notation and, perhaps even more important, to a reliance on geometric methods out of a misguided belief that these represented the spirit of Newton' (Koppelman (1971), pp. 155–6).

The difference between Newton's and Leibniz's notation has been given too much importance. Even though there are some reasons for preferring the differential notation, it is certain that the progress of the calculus of fluxions was not dependent upon the choice between the dots and the d's. Indeed the fluxional notation is still successfully used in mechanics to express the derivatives as a function of time.

Another commonplace misinterpretation is that British mathematicians used geometrical methods. It is not clear to me how the researches of Stirling on interpolation, or of Taylor on finite differences, the second book of Maclaurin's *Treatise of Fluxions* (1742), the work of Simpson on physical astronomy and geodesy, the results of Landen on infinite series and elliptic integrals, and those of Waring on fluxional equations could be defined as geometrical. Many British mathematicians consciously departed from the geometrical methods of the *Principia*, and they did so with different motivations and different results.

The current account of the decline of the calculus of fluxions also includes sociological discussions. It is maintained that the practical bent of a country dominated by the industrial revolution together with the chauvinistic isolation of British scientists caused the stagnation of mathematics. However, many British scientists cultivated a deep interest in pure research, for instance in pure geometry or cosmology, and in Great Britain there was a considerable interest in mathematics as the many 'philomaths' mathematical serials show. The existence of a chauvinistic myth for the *Philosophia Britannica*[4] is undeniable, but this does not imply that there was a total separation between continental and British scientists. For instance, continental and British astronomers were in close contact. Furthermore, the theory of the 'golden isolation' of the fluxionists does not explain why there should be so many letters from continental mathematicians in the correspondence of Stirling and Maclaurin and why there existed several translations from continental mathematical works into English using Newton's notation.[5]

It is disappointing that the only work devoted to the eighteenth-century British calculus, Florian Cajori's *A History of the Conceptions of Limits and Fluxions in Great Britain from Newton to Woodhouse* (1919), restates the usual account. For instance, on p. 254 Cajori simply says that 'Newton's notation was poor and Leibniz's philosophy of the calculus was poor', a statement which historians of Leibniz's mathematics would not easily accept; while on p. 279 we find that 'the doctrine of fluxions was so closely associated with geometry, to the neglect of analysis, that, apparently, certain British writers held the view that fluxions were a branch of geometry'.

Furthermore, Cajori is interested only in the definitions of the term 'fluxion'. Since these definitions did not change very much during a whole century and were generally unsatisfactory from a modern point of view, he takes it as an argument in favour of the thesis of the decline of the British calculus. Cajori's quotations are invariably taken from introductions and prefaces of treatises on fluxions. The reader is left without any information about the authors, the length and contents of their works, and the purposes for which they were written.

Thanks to the recent works of Schneider (1968), Gowing (1983) and Feigenbaum (1985) we have acquired a very good knowledge of de Moivre, Cotes and Taylor. However, it seemed to me necessary to study the whole period from Newton's work to the reform of the calculus in the early nineteenth century. I will offer a general survey of the development of the calculus in Great Britain; I will not consider therefore the impact of the Newtonian calculus on continental mathematics. I will try to concentrate especially on aspects which are not covered in other works. Whenever it is possible, I will refer the reader to studies which cover specific subjects or authors. First of all, I will take for granted a knowledge of Newton's mathematical work, which has been extensively and masterfully studied by Whiteside in his well-known edition of Newton's mathematical papers. Other works which have been useful are: Tweedie (1922) and Krieger (1968) on Stirling; Eagles (1977a) and (1977b) on David Gregory; Clarke (1929) on Simpson; Tweedie (1915), Turnbull (1951) and Scott (1971) on Maclaurin; Grattan-Guinness (1969) and Giorello (1985) on Berkeley; Trail (1812) on Simson; Chasles (1875) on Simson, Stewart and Maclaurin; Smith (1980) on Bayes; and Bos (1974) on the differential calculus. The *Dictionary of National Biography*, the *Dictionary of Scientific Biography* and E.G.R. Taylor's *Mathematical Practitioners* (1954) and (1966), have been indispensable tools in this work. However, the most important source of information on the lives and works of British mathematicians is the monumental P.J. and R.V. Wallis's *Biobibliography of British Mathematics and its Applications* (1986), which I have been able to use at the final stage of my research.

The Overture is devoted to Newton's published work on the calculus of fluxions. Its aim is to present the fundamental elements of Newton's calculus.

The first chapter is concerned with the early diffusion of the calculus of fluxions from 1700 to 1730. The first attempts to popularize the Newtonian calculus were carried out by quite obscure mathematics teachers and itinerant lecturers, such as Charles Hayes, John Harris, Humphry Ditton and Edmund Stone. An analysis of their textbooks shows that they were influenced by the Leibnizian as well as by the Newtonian tradition. The second chapter deals with the research in pure mathematics done by the early Newtonians. Of particular importance are the researches on integration by Roger Cotes, on finite differences by James Stirling and Brook Taylor, and on higher ordered curves by Colin Maclaurin. It seems that early Newtonians, rather than researching the calculus of fluxions, developed related theories, especially the theory of series. In the third chapter space is given to the controversy on the foundations of the calculus originated by Berkeley's *Analyst* (1734). The most authoritative answer to Berkeley was in Maclaurin's *Treatise of Fluxions* (1742): the true manifesto of the fluxionists.

The fourth, fifth and sixth chapters are devoted to the middle period of the fluxional school, roughly from 1736 to 1785. The production of new treatises and the improvements in the applications of the calculus of fluxions occupy, respectively, the fourth and fifth chapters. Particular importance is given to Maclaurin's and Simpson's study on the attraction of ellipsoids. The sixth chapter is concerned with the attempts made by some British mathematicians to develop new techniques in the calculus. A comparison with the progress on the continent shows that the Leibnizian calculus developed into a new form: it became an analytical tool dealing with multivariate functions. Interest in the work of continentals stimulated Thomas Simpson, John Landen and Edward Waring. However, they largely failed to understand the novelty of the analytical techniques of the continentals.

Chapters 7, 8 and 9 are devoted to the reform of the calculus which took place in the period 1775–1820. Four schools of reformers were involved, geographically situated in Edinburgh, the military schools of Woolwich and Sandhurst, Cambridge and Dublin. This part of the book is based on completely unknown material, the contribution of two generations of

British mathematicians having been ignored by historians. It is argued that the work of these mathematicians, including Charles Hutton, John Playfair, James Ivory, William Wallace, John Brinkley and Robert Woodhouse, laid the foundations for the resurrection of British mathematics in the first half of the nineteenth century.

In Appendix A I have grouped the tables which give information on the content, and in appendix B I have given the prices of some textbooks on fluxions. Appendix C lists the Chairs of Mathematics in Cambridge, Oxford, Edinburgh, Glasgow, St Andrews, Aberdeen and Dublin, and Appendix D gives information on the teaching of mathematics in the military schools at Woolwich, Sandhurst and Portsmouth. A subject index of the primary literature is given in Appendix E, and a list of the manuscript sources used is in Appendix F.

After these appendices the reader will find the endnotes, the general bibliography and the index.

This book therefore covers more than a century. From necessity I have been extremely selective in the analysis and discussion of the works connected with the development of the fluxional calculus. I have chosen those which appeared to me more exemplary of the level of research and style of a determinate mathematician or group of mathematicians. In compiling the bibliography, on the other hand, I have tried to be as complete as possible. I hope that my work will be useful as a first survey and historical assessment of the contributions (and failures) of British mathematicians in the eighteenth century.

This book is an improved version of my Ph.D. thesis submitted in June 1987 to the Council for National Academic Awards. A three year scholarship of the Italian Ministero della Pubblica Istruzione (D.M. 27.1.83) and a two year appointment as part-time research assistant at Middlesex Polytechnic (UK) provided the financial support which allowed me to complete my doctorate. My interest in Newton's calculus originates from the thesis I wrote in 1981–2 under the supervision of Prof. Corrado Mangione at Milan University. I was then encouraged by several friends, among whom the most encouraging was Giulio Giorello, to pursue and extend my research. Next I must mention Allan Findlay and Ivor Grattan-Guinness, the supervisors of my Ph.D. thesis during the years 1984 to 1987. Ivor followed my every step and gave to me all his encyclopaedic assistance: I owe very much to his competence, but especially to his friendship. I would like also to thank Eric Aiton, the external examiner of my Ph.D., who directed my attention on many points which needed amendments and additions. During several stays in Cambridge I received

xii   INTRODUCTION

advice from Michael Hoskin and Simon Schaffer. In Cambridge I had the
privilege of meeting Tom Whiteside, the great authority on Newton's
mathematics, who, with great kindness and generosity, criticized several
drafts of my book. I also owe a great deal to Roger Bray for giving me
important information on the military schools, to Jennifer Carter for her
kind letters on Aberdeen University, to Marco Panza for sending me early
drafts of his (1989), and to Eric Sageng for his advice on James Gregory
and Colin Maclaurin. Luca Bianchi, Umberto Bottazzini, Michele Di
Francesco, Massimo Galuzzi and Angelo Guerraggio have been important
in the progress of my research in several ways. I would like also to thank
the librarians of Cambridge University Library (most notably Stephen
Lees), the British Library (London), the Senate House Library (London),
and the Royal Society Library (London). Thanking all these people is the
part of my research which gives me the most satisfaction.

OVERTURE: *NEWTON'S PUBLISHED*
WORK ON THE
CALCULUS OF FLUXIONS

WHEN AT the beginning of the eighteenth century the Newton–Leibniz controversy exploded, the great majority of British scientists declared their loyalty to Isaac Newton. In their opinion the calculus of fluxions had been invented long before the differential calculus of Leibniz and had been stolen by the German philosopher. It was also generally maintained that Newton's calculus was more firmly grounded than Leibniz's: not only was it thought that Newton had been the first to invent the calculus, but also that he had laid its scientific foundation. It was not clear, however, which was the genuine form of the calculus of fluxions since it was difficult to derive a coherent idea of it from Newton's work.

Newton's first published work on fluxions, 'De quadratura' (1704c), appeared in 1704 as an appendix to *Opticks* (1704a). This was followed in 1711 by the 'De analysi' (1711b), a tract written in c. 1668/9.[1] And then, of course, there were the *Principia* (1687) with their huge apparatus of mathematical techniques. It was not until 1736 that there appeared in English, with the title *The Method of Fluxions and Infinite Series*, the translation of a long and comprehensive treatise which Newton had composed in 1671.[2] All the Newtonians agreed that the calculus of fluxions was contained in these works. However, in the *Principia* there was no proper algorithm of fluxions, even though series and some algebraical manipulations with 'moments' occurred, especially in the second book. The two treatises of 1704 and 1711 were in sharp contrast concerning the foundation of the calculus, since in the former Newton tried to avoid the use of infinitesimals, while in the latter infinitely small quantities were freely employed. If Newton was the true inventor of the calculus, where did he present the genuine form of it?

Newton himself tried to give an answer in the anonymous 'Account' (1715) to the *Commercium Epistolicum* which appeared in the *Philosophical*

Transactions for the year 1715. According to him, the whole controversy with Leibniz

relates to a general Method of resolving finite Equations into infinite ones, and applying these Equations, both finite and infinite, to the Solution of Problems by the Method of Fluxions and Moments. (Newton (1715), p. 173)

Newton made it clear that his theory was divided into two parts: the first dealt with series, while the second consisted in the application of the method of fluxions and moments to infinite series or to finite algebraical formulae. In fact in 'De quadratura' (1704c) he explained how the two parts could be used together. Given an expression in which there occurred the fluxions $\dot{y}$ and $\dot{x}$ of unknown quantities, he could manipulate the expression in order to obtain a power series of $\dot{y}/\dot{x}$. He was then able to operate on the series applying simple methods of integration and determine y.[3] In modern terms, he integrated first order differential equations, expanding the derivative of the unknown function and integrating term by term. Indeed Newton's method allowed the integration of differential equations that were quite complicated for late seventeenth-century mathematicians.

For instance, he considered:[4]

$$\frac{\dot{x}}{\dot{y}} = \frac{1}{2}y - 4y^2 + 2yx^{\frac{1}{2}} - \frac{4}{5}x^2 + 7y^{\frac{5}{2}} + 2y^3; \; x(0) = 0.$$

His procedure can be rewritten as follows:

$$\text{if } y \simeq 0, \text{ then } \frac{\dot{x}}{\dot{y}} \simeq \frac{1}{2}y.$$

Therefore,

$$x = \frac{1}{4}y^2, x^2 = \frac{1}{16}y^4, x^{\frac{1}{2}} = \frac{1}{2}y.$$

Substituting:

$$\frac{\dot{x}}{\dot{y}} = \frac{1}{2}y - 3y^2 + 7y^{\frac{5}{2}} + 2y^3 - \frac{1}{20}y^4, \frac{\dot{x}}{\dot{y}} \simeq \frac{1}{2}y - 3y^2.$$

Therefore,

$$x = \frac{1}{4}y^2 - y^3,$$

$$x^2 = \frac{1}{16}y^4 - \frac{1}{2}y^5 + y^6,$$

$$x^{\frac{1}{2}} = \frac{1}{2}y - y^2 + \ldots.$$

Substituting:

$$\frac{\dot{x}}{\dot{y}} = \frac{1}{2}y - 3y^2 + 7y^{\frac{5}{2}} - \frac{1}{20}y^4 + \frac{2}{5}y^5 + \dots,$$

$$\frac{\dot{x}}{\dot{y}} \simeq \frac{1}{2}y - 3y^2 + 7y^{\frac{5}{2}}.$$

Therefore,

$$x = \frac{1}{4}y^2 - y^3 + 2y^{\frac{7}{2}} + \dots.$$

The interest in series representations was brought about by the possibility of treating transcendental functions as algebraic functions. In fact, power series were treated as infinite polynomials: considerations on convergence emerged only at a very intuitive level.

When in the 'Account' Newton came to characterize the 'method of fluxions and moments' he presented side by side two different explanations of it. In the first he used moments, 'momentaneous Increases, or infinitely small Parts of the Abscissa and Area, generated in Moments of Time' (Newton (1715), p. 178). In modern notation the moments can be represented as $y'(t)dt, x'(t)dt$, etc. Newton used $\dot{y}o, \dot{x}o$, where $\dot{x}$ and $\dot{y}$ are the fluxions, or instantaneous velocities, of x and y and o is an infinitely small interval of time. Apart from their kinematical generation, Newton's moments are equivalent to Leibniz's differentials. Therefore in what follows note that the Leibnizian dx corresponds to Newton's $\dot{x}o$, where $\dot{x} = dx/dt$, and o is an infinitesimal interval of time dt.

But Newton also presented another approach to the calculus and this was, according to him, a rigorous and genuine approach.

When he [Newton] is demonstrating any Proposition he uses the Letter o for a finite Moment of Time, or of its Exponent, or of any Quantity flowing uniformly, and performs the whole Calculation by the Geometry of the Ancients in finite Figures or Schemes without any Approximation: and so soon as the Calculation is at an End, and the Equation is reduced, he supposes that the Moment o decreases in infinitum and vanishes. But when he is not demonstrating but only investigating a Proposition, for making Dispatch he supposes the Moment o to be infinitely little, and forebears to write it down, and uses all manner of Approximations which he conceives will produce no Error in the Conclusion. (Newton (1715), p. 179)

As Bos (1974) has shown, behind these two different presentations of the calculus there are two different conceptions of continuous magnitude. Let us take for example a curve C (see fig. 1) and the variables associated with it: the abscissa x, the ordinate y, the subtangent t, the tangent T, the arclength s, the area Q, the normal n. In the calculus of moments or

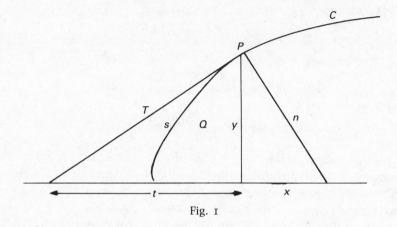

Fig. 1

differentials these continuous magnitudes can be analysed in terms of their infinitesimal components.[5] In fact, the variables are supposed to increase or decrease by infinitesimal steps y, y', y'', etc. Therefore we can associate with this progression a progression of differentials defined as $\mathrm{d}y = y' - y$, $\mathrm{d}y' = y'' - y'$, etc. Similarly, second order differentials are defined as $\mathrm{d}^2y = \mathrm{d}y' - \mathrm{d}y$, $\mathrm{d}^2y' = \mathrm{d}y'' - \mathrm{d}y'$, etc. The required information on the curve C will determine how deep the analysis must be. For instance, to determine the subtangent it is sufficient to consider first order differentials, while to determine the radius of curvature it is necessary to analyse second order differentials.

There are three characteristics of the calculus of moments which should be noted.

(1)   *The arbitrariness in the choice of the progressions of the variables.* In the analysis of the curve C it has to be established which, if any, of the variables satisfies the condition $\dot{w}o = constant$. For instance, it can be assumed that $\dot{s}o = 1$. In this case the curve is subdivided into equal infinitesimal arcs, and this partition on the curve will determine unequal partitions of the other variables, for instance x and y. Any of the variables can be supposed to be subdivided so as to satisfy the condition $\dot{w}o = constant$, there is no variable which takes precedence. A more general view can be taken and this restrictive condition will not hold for any of the variables.

(2)   *A variable and its differentials are dimensionally homogeneous.* In fact, if, for instance, Q has dimension 2, $\dot{Q}o$, $\ddot{Q}o^2$, $\dddot{Q}o^3$, etc. will have the same dimension, they will be infinitely small areas.

(3)   *The Archimedean principle does not hold universally.* In the calculus of

moments it may be possible to have two homogeneous magnitudes a and b, such that $a < b$ and there is no n such that $an > b$. The magnitudes for which Archimedes's principle holds belong to the same class of infinity.

In the second approach to the calculus, which we may call the calculus of limits, magnitudes are not analysed in terms of their infinitesimal components. In this case the geometrical magnitudes defined, for instance, by curve C in fig. 1, must be considered to be in the final state of a process. A possible choice is to conceive this process as a kinematical one: using a modern terminology one might say that the magnitudes grow and decrease as 'functions of time'. In fact Newton called the magnitudes 'fluent quantities' and their rates of increase 'fluxions'. In the final state, at a certain determinate instant, some quantities may disappear simultaneously, and, as is well known, in geometry and mechanics it is important to establish the limit to which the ratios and sums of these 'vanishing quantities' approach. Consequently, Newton developed an intuitive theory of limits: the 'method of prime and ultimate ratios'. Also in this second approach to the calculus a choice has to be made on the progression of the variables. For instance, it can be assumed that $\dot{x} = constant$, or, as Newton said, 'x flows uniformly'. Again no variable must be privileged; there is no need to suppose that any of the variables flows with a constant velocity. However, characteristics (2) and (3) of the calculus of moments have no place in the calculus of limits: continuous magnitude is not analysed in terms of homogeneous components belonging to an inferior class of infinity. In the calculus of limits $\dot{y}$ is not an infinitesimal but a finite rate of change of y.

Newton's aim in the 'Account' was to defend his position against the Leibnizians. He tried to show that the calculus of fluxions had been invented in all its details (notation, methodology, theorems) before 1684 and that it was superior to Leibniz's differential calculus. But Newton's presentation of the calculus changed from the 1660s to the 1690s, while his notation $\dot{x}$, $\dot{y}$, $\dot{z}$ was employed by him only in the early 1690s. In his early writings on the calculus, for instance in the 'De analysi' (1711b), Newton's use of infinitesimals was very similar to Leibniz's differential calculus. Furthermore, in those early years Newton was employing a notation clearly inferior to Leibniz's (Newton used small a, b and c to represent the infinitely small increments of A, B and C). These facts were hidden by Newton in the 'Account', which unfortunately was to be considered by his immediate followers, to whom the author of this anonymous review of the *Commercium Epistolicum* was not a mystery, as the last word of Newton on the calculus of fluxions. As a matter of fact in Newton's mathematical work it is possible to distinguish four different alternatives.

The propositions in the fluxional calculus could be expressed as:

(1)   using differentials as infinitely small components of finite magnitudes;

(2)   using moments as infinitely small components generated by motion in an infinitely small interval of time;

(3)   using finite variable quantities and their rates of change; and

(4)   using flowing quantities and their velocities (or fluxions).

In cases (1) and (2) the proofs are based on a principle of cancellation according to which if a is infinitely small in comparison with A, then a can be eliminated so that $A + a = A$. In cases (3) and (4) the proofs are based on an intuitive theory of limits. The limits are taken considering the quantities in the final state of a process of successive approximations (in (3)) or in the final state of a kinematical process (in (4)). Here a is finite, but, as the process goes on, a becomes smaller than any given difference, therefore in the limit $A + a = A$.

In his early writings, most notably the 'De analysi' and the 'De methodis' (published later as (1711b) and (1736)), Newton employed the principle of cancellation of higher order 'differentials' (one can use here Leibniz's terminology since Newton's technique is equivalent to Leibniz's $x + dx = x$). Later, in the *Principia* (1687) and 'De quadratura' (1704c), Newton developed a theory of limits, which was understood by him as the correct foundation of the calculus, while his early differentialist methods were to be considered as just abbreviations.[6] It is especially in the preparation of 'De quadratura' (1704c), i.e. in the period 1693–1704, that Newton conceived the theory of limits as a means of banishing infinitely small quantities. As we have seen, these developments were hidden during the Newton–Leibniz controversy. For instance, in the 'Account', just after a presentation of his theory of limits, as it was developed in 'De quadratura' (1704c), Newton could continue by saying that the calculus of fluxions was superior because in it 'there is but one infinitely little Quantity represented by a symbol, the symbol o'. This sentence could be certainly employed as a presentation of the method of moments developed in the 'De methodis' of 1671, but was in complete contradiction with the method of limits of 'De quadratura' (1704c). Newton left to his followers a wealth of theorems and results achieved with the calculus of fluxions, but he left also a rather confused presentation of the methods and of the concepts which had to be employed. As we will see in chapter 3, the debate on the foundations of the fluxional calculus, a debate which occupied eighteenth-century British mathematicians so much, is, to a great extent, the history of different interpretations of the mathematical work of Isaac Newton.

PART I

THE EARLY PERIOD

1

THE DIFFUSION OF THE CALCULUS
(1700–30)

AT THE beginning of the eighteenth century, Newtonian Natural Philosophy began to spread all over Europe. In Great Britain an astonishing amount of interest was shown in the new science. A great number of lectures were given both in and outside London. The role of the fluxional calculus in this process of reinterpretation and popularization was negligible. Quite understandably, a technical subject did not attract the audiences who gathered to attend the lectures on Natural Philosophy.

However, a group of 'philomaths' was already active in the seventeenth century. They were most likely interested in Newton's mathematical theory; and, when properly instructed, they were probably able to develop the calculus of fluxions. But in the later years of the seventeenth century the Newtonian calculus could only have been known to Newton's correspondents, such as Collins, Oldenburg and Wallis. Manuscript copies of parts of Newton's work on fluxions were in circulation. The few who were allowed to glance at them, such as Craige, Fatio and David Gregory, were able to grasp just a fraction of Newton's achievements.

Needless to say, the first persons to try and systematize this information in the form of a treatise had to use as their sources the *Acta Eruditorum* and L'Hospital's *Analyse* (1696). Particularly remarkable were the attempts of Cheyne (1703) and Hayes (1704).

In later years the calculus enjoyed a time of great popularity thanks to the priority controversy with Leibniz. I will not even try to retell the story of this famous dispute, which has already been described in all its details and implications by Hofmann (1943), Fleckenstein (1956), Whiteside in Newton (1967–81), VIII, Scriba (1969), and Hall (1980). What concerns us is the fact that the quarrel with the Leibnizians caused the publication of Newton's mathematical tracts (1711a) and of parts of his mathematical correspondence as Collins (1713). Newton's countryfellows were able to read and digest the lesson.[1]

This lesson, however, arrived late in the universities. In Oxford the splendid group formed by Gregory, Keill, Halley and Bradley did very little to promote the study of mathematics. In Cambridge, on the other hand, we find Saunderson teaching fluxions as fourth Lucasian Professor. Glasgow with Simson and Edinburgh with Maclaurin also played an important role in the diffusion of fluxions in the British universities.

1.1 Early initiates

The beginnings of the diffusion of the fluxional calculus are even more uncertain and confused in comparison with the diffusion of Newton's theory of gravitation and his optics. This was due mainly to the well-known reluctance of Newton to publish his mathematical work: only his few acquaintances and correspondents, such as Wallis, Oldenburg and Collins, were initiated into the calculus of fluxions. In fact Fatio de Duillier (1664–1753), John Craige (d.1731) and David Gregory (1659–1708), the first persons to try and develop the calculus of fluxions, had to consult Newton personally on his not yet published 'new analysis'.

Fatio was for a period one of Newton's *protégés* and was allowed, in the winter of 1691–2, to read parts of the preparatory manuscripts of 'De quadratura' (1704c). Not at all a mediocre mathematician, Fatio is the author of *Lineae Brevissimi Descensus Investigatio Geometrica* (1699) in which he presents his own solution of the well-known brachistocrone problem. However, his work is more famous for having opened the priority controversy by stating, clearly and for the first time, that Leibniz had plagiarized Newton's calculus.

John Craig(e), a Scot who resided for some time in Cambridge, also enjoyed a privileged relationship with Newton. He was able to study in 1685 parts of the 'De analysi' and of the 'De methodis' (which were later published as Newton (1711b) and Newton (1736)). Craig(e) published in (1685) the first British work on quadratures: the *Methodus Figurarum Lineis Rectis & Curvis Comprehensarum Quadraturas Determinandi*. It is ironic that this work and his subsequent *Tractatus Mathematicus de Figurarum Curvilinearum Quadraturis et Locis Geometricis* (1693) were written in differential notation. In these years prior to 1704, Newton was able to act as a personal guide for the few lucky ones, but the published sources for the early British 'fluxionists' were the *Acta Eruditorum* and, from 1696, L'Hospital's *Analyse des Infiniments Petits* (1696). In fact all Craig(e)'s mathematical work is fully within the tradition of the Leibnizian calculus. He described his first work in the *Philosophical Transactions* for the year 1686 as independent from Newton's. He mentioned the researches of

Descartes, Fermat, Sluse, Barrow, Wallis, Tschirnaus and Leibniz, but added:

none has attempted to invert this problem generally, that is, having the Tangent to find the Curve Line whose Tangent it is. (Craig(e) (1688a), p. 185)

So Craig(e) considered his (1685) as the first work on integration. After reading Leibniz's 'Nova methodus' (1684), he tried to reformulate in differential notation some theorems of Barrow on integration by substitution of variables. His idea consisted in expressing the change of variable as a multinomial with undetermined coefficients.

The case of David Gregory illustrates very well how difficult it was in this period to obtain information on the calculus of fluxions.[2] He was the first person to try and write, between 1694 and 1695, a small tract on fluxions. He was able to base his work on the few pages of John Wallis (1693), pp. 390-6, in which passages of Newton's manuscript 'De quadratura' were included.[3] He also had a manuscript copy, taken in 1685 by Craige, of the first parts of Newton's 'De methodis' which do not deal with fluxions but with series.[4] Further information could be gained only by visiting Newton and displaying due deference. In May 1694 Gregory went to Cambridge and was allowed by Newton to take copies of the 'truncated' version of 'De quadratura'. Halley was more fortunate and was able to take the manuscript itself to London. Examples of application of the calculus were to be found later in the *Philosophical Transactions*,[5] but a comprehensive exposition of the calculus of fluxions was still lacking.

1.2 Textbook writers

We can consider John Harris's *New Short Treatise of Algebra* (1702) as the first published introductory presentation of the calculus of fluxions. Harris wrote:

There being nothing published on this Subject [the calculus of fluxions] in our own language, and yet the vast use of this *Method of Investigation* being as conspicuous, as it is wonderful: I thought it proper to give a short account of it here. (Harris (1702), 2nd edn., p. 133)

Harris devoted only twenty pages of his *Algebra* to fluxions and (in the second edition of 1705) he referred the enthusiast to the following texts:

he [the reader] will find sufficient Satisfaction, by perusing the Authors above mentioned, vz Newton, Wallis, Niewentiit, Carre, Leibnitz (in the Act.Eruditor.Lipsiae) and especially the Marquis L'Hospital, his excellent Analyse des Infiniment Petits. (Harris (1702), 2nd edn., p. 149)[6]

John Harris (1666–1719) was one of the most influential mathematics teachers in London, and one of the first to adopt Newton's natural philosophy. He was elected Fellow of the Royal Society in 1696, and for a short period (1709–10) he was Secretary, but he suffered Hans Sloane's opposition. In 1698 he delivered the Boyle lectures in St Paul's Cathedral. As a teacher, Harris was never connected with any institution: in around 1698 he began to give free public lectures on mathematics at the Marine Coffee-House in Birchin Lane. We read in a note 'To the Reader' in Harris's *Algebra*:

This small tract of that Admirable Science, Algebra, was written primarily for the Use of my Auditors at the Publick Mathematick Lecture, which was set up at the Marine Coffee-House in Birchin-Lane, intirely for the Publick Good, by the Generous Charles Cox, Esq.; Member of Parliament for the Burgh of Southwark. (Harris (1702), 2nd edn.)

Harris's most successful work was the *Lexicon Technicum* (1704, 1710) a huge scientific dictionary in two volumes. Harris's dictionary rendered obsolete Joseph Raphson's *Mathematical Dictionary* (1702) which was derived from Ozanam's *Dictionnaire Mathématique* (1691). Harris's work included articles on series, algebraic equations, trigonometry and conics. For our purpose it is important to note that in the second volume (1710) there was an English translation of Newton's *De quadratura* (1704c) under the title 'Quadrature of curves'. However, as we will see in chapter 3, Harris did not feel the need to reconcile it with the article 'Fluxions' in the first volume which was based on completely different principles.

Another complete work on mathematics which included a treatment of fluxions was *Synopsis Palmariorum Matheseos* (Jones, 1706). The author, William Jones (1675–1749), was a friend of John Harris, probably one of his pupils. He also established himself as a teacher of mathematics. Jones was tutor to George Parker (later second Earl of Macclesfield and President of the Royal Society). He is also well known as the editor of Newton's *Analysis per Quantitatum, Series, Fluxiones, ac Differentias* (1711a) and as a member of the committee established to decide on the Newton–Leibniz controversy. He was elected Fellow of the Royal Society in 1711, and later became Vice-President. Even though Jones's *Synopsis* (1706) went through only one edition it was very well known and often cited.[7] It was an advanced textbook which covered arithmetic and algebra in the first part, and in the second ('containing the principles of Geometry') conic sections, plane and spheric trigonometry, mechanics and optics. Only a few pages were strictly devoted to the calculus of fluxions: Jones presented Newton's notation, gave the rules of differentiation of the elementary functions and applied the 'inverse method' to power series. More interesting was the section dealing with mechanics where Jones treated some propositions of the *Principia*. It was probably this section that rendered Jones's *Synopsis* so useful at the beginning of the century, in a period in which Newton's major work was hard to understand even for the best mathematicians.

The iatro-mechanist[8] George Cheyne (1671–1743) tried with his *Fluxionum Methodus Inversa* (1703) to write a treatise completely devoted to the fluxional calculus, which, as we have seen, was at that time still lacking. Even though he was not a mathematician, he was induced to attempt this task by his protector, Archibald Pitcairne. The results was, as an expert judge like Whiteside puts it,

a competent and comprehensive survey of recent developments in the field of 'inverse fluxions' not merely in Britain, at the hands of Newton, David Gregory and John Craige, but also by Leibniz and Johann Bernoulli on the Continent, and drew the assemblage together and systematized it with proofs and elaborations of Cheyne's own contrivance (Newton (1967–81), VIII, pp. 17–18).

This work was, however, to arouse the anger of Newton who felt, without any justification it would seem, the threat of being plagiarized by a man of so little mathematical skill. Abraham de Moivre was instructed to attack Cheyne, and he did so with vehemence in his *Animadversiones in D. Georgii Cheynaei Tractatum de Fluxionum Methodo Inversa* (1704b).[9] In the meantime Newton felt that it was time to publish his treatise on quadratures written in 1693; he appended it with the title 'Tractatus de quadratura curvarum' (1704c) to the *Opticks* (1704a). The harsh reception of Cheyne's work certainly damaged its popularity. Furthermore, it was written in a somewhat confused style. As a result, it exerted a minimal influence on the British calculus. But other works were about to appear which could meet the increasing demand for information on the 'new analysis' of Newton.

These were Hayes's *Treatise of Fluxions* (1704) and Ditton's *Institution of Fluxions* (1706). The aim of Hayes and Ditton was clearly to write treatises to be used as introductions accessible to readers who were unacquainted with the calculus. For instance, Hayes wrote:

As to the ensuing Treatise, the Author has been assur'd that there are in England as many Lovers of the Mathematicks as in any part of the World; that multitudes of excellent Judgements and natural Parts, merely for want of a competent Knowledge in other Languages, have hitherto been deprived of the Opportunities of improving them, to the great disadvantage of the most *Flourishing Island* in the *World*; that in other Nations the best pieces of Learning are written in their own mother Tongues, for the good of their Country which we seem purposely to slight, seeking a little empty applause by writing in a Language not easily attain'd, as if the Knowledge of *things* and *words* had a necessary dependance on each other; and in a word, that such a Treatise was wanting in the English Tongue, as should contain a full and plain account of the best Methods, the most celebrated Geometers of our Age have made use of in their wonderfull Discoveries; and which would put it in the Power of every industrious Person to make use of those parts which *God* and Nature has bestow'd upon him to the best purposes: These, he says, were the principal motives that induced him to this difficult undertaking, and he hopes the sincerity of his design will at least merit favourable Censure from the World. He knows there are persons better qualified for such an undertaking, but none appearing, hopes his forwardness to serve the Publick will be no objection against him. (Hayes (1704), from the first page of the 'Preface to the Reader')[10]

Charles Hayes (1678–1760) was not a teacher of mathematics. However, he had a good knowledge of mathematics as he is described as expert in geography and cartography: it is probably because of his skill in these fields that he was chosen to be deputy governor of the Royal African Company. Hayes (1704) was an outstanding achievement for a self-taught man. In the Preface he writes that he asked the advice of John Harris, but he clearly outdid his supervisor. Hayes's treatise covered in more than 300 pages all the known areas of the early eighteenth-century calculus: from the brachistocrone and the study of radii of curvature to the quadrature of 'mechanical' curves. But this treatise was published just before Newton's masterpiece (1704c), which certainly nullified the usefulness of Hayes's laudable effort.

Humphry Ditton (1675–1715), a teacher of mathematics in London, was appointed, through Newton's influence, Master of a *New* Mathematical School at Christ's Hospital.[11] His *Institution of Fluxions* (1706) was more detailed and accurate on foundations but less advanced in contents than Hayes (1704). Ditton's treatise was not written specifically for Christ's Hospital since only 'forty poore boys' received an education in the mathematical school. We must assume, therefore, that Ditton too wrote in order to meet the demand from the many 'lovers of mathematics'.

Other publications on fluxions were the *Commercium Epistolicum* (Collins (1713)) and Joseph Raphson's *History of Fluxions* (1715). The primary aim of these works was to establish Newton's priority in the invention of the calculus, and indeed they cannot be defined as textbooks (see note 1 of this chapter). However, some information on the calculus could be derived from them. The former (Collins (1713)) was a result of the Committee of the Royal Society set up to decide on the Newton–Leibniz controversy and most notably included the two famous 1676 *epistolae* of Newton to Leibniz and Newton's 'De analysi' (1711b). The latter consisted of a rather confused series of quotations from Leibniz's 'Nova methodus' (1684),

Newton's 'De quadratura' (1704c), Cheyne's *Fluxionum Methodus Inversa* (1703), and Craig(e)'s paper (1698a).

It seems that in the first three decades of the century there was a very limited market for treatises on the calculus of fluxions. Hayes (1704) ran only one edition, while Ditton was republished in 1726. Newton's collection of mathematical tracts (1711a) clearly exhausted the demand. We have to wait twenty-four years for another introductory textbook on fluxions, Stone's *The Method of Fluxions* (1730). On the other hand, in this period a great number of popular introductions to Newton's astronomy, mechanics and optics were published. In these works experiments rather than mathematical proofs were dominant. As we will see in chapter 4, treatises on the fluxional calculus began to flourish in the late 1730s.

Edmund Stone (1695–1768) was another writer of mathematical texts who, like Hayes, had no connection with any educational institution. He was the son of a gardener of John Campbell, second Duke of Argyll at Inverary. Stone published in (1726) *A New Mathematical Dictionary*, a shorter and less expensive alternative to Harris (1704, 1710). But his most substantial work was *The Method of Fluxions* (1730) published in two volumes in 1730.[12]

The first volume consisted of a translation of L'Hospital's *Analyse des Infiniment Petits* (1696) in which Stone simply put in Newton's dots in place of Leibniz's d's. Since L'Hospital had confined his treatise to the differential calculus, in the second volume Stone provided a treatment of the integral calculus (or, better, of the 'inverse method of fluxions'). In 1735 Stone's second volume was translated into French. Stone (1730) is clearly different from Hayes (1704) and Ditton (1706). Hayes (1704), which is the more complete, after a very short presentation of the rules of the 'Algorithm or Arithmetic of Fluxions' immediately moves on to show in the following 300 pages the application of the calculus to a variety of geometrical and mechanical problems (finding tangents, areas, maxima and minima, caustics, centres of gravity, percussion and oscillation, plus a treatment of central forces).[13] Furthermore, even though there were quite a lot of examples of applications of the 'inverse method of fluxions', the mathematical treatment of the rules of integration was almost entirely absent. By translating L'Hospital's *Analyse des Infiniment Petits* (1696), Stone offered to the English reader a systematic and much more analytic treatise on the direct method, with applications only to geometric problems: it was not, like Hayes (1704) and Ditton (1706), a collection of problems with their solutions, but a methodic treatise which included only a few problems and considered them at a quite advanced level (for instance sections VI and VII on caustics). The second volume is not a masterpiece, but it has the merit of including some tables of integrals ('forms of fluents') taken from Cotes's *Harmonia Mensurarum* (1722). However, it does not seem that Stone (1730) enjoyed great popularity since it was superseded by a number of more up-to-date textbooks published in the 1730s and 1740s. Furthermore, the well-known *Analyst*'s controversy provoked by Berkeley (1734) rendered the community of British mathematicians too sensitive to the problem of foundations for them to accept the strictly infinitesimalist treatise of L'Hospital.

1.3 The teaching of the calculus in the universities

The paucity of treatises on fluxions published in the first three decades of the eighteenth century is certainly the result of the predictable difficulties that even the most enthusiastic 'learned men' found in approaching the new analysis. However, the few Newtonians who acted on behalf of Isaac Newton in the universities of England and Scotland might have been interested in promoting the study of fluxions to a small audience: so small as not to justify the cost of a publication. We should note here that in the universities public lectures were given, open not only to the young (sometimes twelve years old) students or to members of the university, but also to groups of adults interested in science. This is true of Keill's, Gregory's and Bradley's lectures in Oxford, as well as Cotes's, Whiston's and Saunderson's lectures in Cambridge. A closer look at the lectures on mathematics, natural philosophy and astronomy in vogue in the universities and at some minor sources will give us further information on the place occupied by the calculus of fluxions in the diffusion of Newtonian science in the early eighteenth century.

It must be stated at the outset that our knowledge of what was really taught in a course of lectures and of what was really read in a treatise is based on quite uncertain data. Sometimes it is even difficult to understand whether lectures were given at all. In fact, we have to distinguish between the evident intentions of the writer or lecturer and the more hidden motivations of his listeners and readers. The historian here has to face the embarrassing truths that not all the subscribers were enthusiastic readers and that they had different motivations (they were friends of the author, they shared his political views, they followed a fashion, etc.); that not all the deposited lectures were actually read and that very often a writer of an introductory text or a teacher addressed himself to a public (e.g. the 'many lovers of science') which might not exist.

David Gregory's *legenda* for his students at Oxford, which are quoted in Eagles (1977a), pp. 134–41, are certainly unrealistic.[14] Gregory, who was

Savilian Professor of Astronomy at Oxford from 1691/2 to 1708, included in the *curriculum studiorum* Euclid's *Elements*, Apollonius's *De Sectione Coni*, Serenus's *De Sectione Cilindri*, Newton's *Principia*, Kepler's *Mysterium Cosmographicum*, *Harmonicae Mundi* and *Astronomia Nova*, and Wallis's *Arithmetica Infinitorum*. It is quite certain that such a course was never followed since it is very well documented that the level of studies in English universities was then very low.

However, Gregory gave some place in his teaching to Newtonian philosophy both in Edinburgh, where he held the Chair of Mathematics from 1683 to 1691, and in Oxford where, as we already know, he wrote a tract on fluxions probably for the use of his (better!) students. John Keill (1671–1721), a friend of Gregory, arrived from Scotland in 1694 and began to lecture on Newtonian philosophy. In 1699 he was employed by Thomas Millington as deputy Professor of Natural Philosophy. His course was published as Keill (1702). From 1709 to 1712 Keill was in New England. His lectures were continued by John T. Desaguliers (1683–1744). In 1712 Keill was elected Savilian Professor of Astronomy and lectured regularly till his death. When Gregory and Keill began their campaign in favour of the establishment of Newton's philosophy in Oxford, they found the Savilian Chair of Geometry still occupied by John Wallis (1616–1703) who was succeeded in 1703 by Edmond Halley (1656–1742). Consequently Oxford became a very interesting centre of diffusion for Newtonianism. Halley, James Bradley (1693–1762) and Nathaniel Bliss (1700–64), Astronomer Royals, respectively, from 1720 to 1742, from 1742 to 1762 and from 1762 to 1764, all held one of the Savilian Chairs. However, even though Gregory, Keill, Halley and Bradley were very well qualified in teaching the calculus of fluxions, there is little evidence that they actually did so.

The manuscript lectures of Gregory do not supply more than a few lectures a year, despite the fact that they cover mechanics, hydrostatics, optics and astronomy (see Eagles (1977a), p. 100). Mathematics is not touched on in these few extant examples, but Gregory's small tract on fluxions, and especially his treatise on geometry, which was published posthumously as (1745), provide sufficient proof that Gregory was interested in teaching the calculus. Eagles, who has analysed Gregory's published and unpublished work in (1977a), draws the conclusion that in his lectures he stressed the utilitarian goals of mathematics: a conclusion which seems plausible if one considers the style of Gregory (1745).

The best source on Newtonianism in early eighteenth-century Oxford are Keill's lectures published as *Introductio ad Veram Physicam* (1702) and as *Introductio ad Veram Astronomiam* (1718). These lectures were also widely diffused outside Oxford: they were translated into English and served as a model for many other treatises on natural philosophy and astronomy. However, the reader will find in Keill's exposition views very far from anything that Newton had ever written: for instance, on the nature of matter or on the force of gravitation. In fact early Newtonians, such as Roger Cotes (1682-1716), William Whiston (1667–1752) and Keill, offered global reinterpretations of the *Principia* which very often went far from the intentions of Newton.

We will not be surprised, then, to find in Keill's lectures a relationship between mathematics and physics which was never envisaged by Newton. Keill argued in favour of the infinite divisibility of matter by 'Arguments taken from Geometry' (Keill (1720), p. 21). He tried to 'demonstrate that all Extension, whether corporeal or incorporeal, was divisible *in infinitum*, or had an infinite Number of Parts' (Keill (1720), p. 26). Keill did not explicitly distinguish between 'geometry' and 'physics' and interpreted the Euclidean proofs on the infinite divisibility of magnitude as applicable to the analysis of the structure of matter. If we can conceive geometrical magnitudes as well as matter as composed of an infinity of infinitely little parts, we can proceed to infinitesimals of the second order. It is here that Keill introduces the term 'fluxio' for the first time in this series of lectures:

since they [the infinitely small parts] are extended, they will be also divisible; not only in two or three, or more Parts, but likewise every one may be divided *in infinitum*. The infinite Number of Parts of an infinitely small Quantity, are wont to be called by the Geometers, Infinitesimals of Infinitesimals, or Fluxions of Fluxions. (Keill (1720), p. 40)

In order to prove the 'existence' of different orders of infinitesimals, Keill employed the reasonings on angles of contact which could be found in the final scholium of section I of Newton's *Principia* (Keill (1720), pp. 43–5). It is interesting to note that the Newtonian calculus appeared, even though in a very strange form, not in Keill's treatment of mechanics or astronomy but in his considerations on matter theory.

Even though Keill was one of the most vehement defenders of Newton in the priority dispute with Leibniz on the invention of the calculus, it seems that he did little to promote the application and the development of the calculus of fluxions. Indeed mathematical methods were employed in some of his lectures (e.g. the Lectio XV of his (1720) on the motion on inclined planes and pendulums, and the Lectiones XXIV and XXV of his (1718) on Kepler's problem), but generally Keill preferred to tackle astronomical and mechanical problems by using the geometry of the

Principia. In other cases Keill did not even attempt to give a mathematical proof for every proposition, but rather he described the Universe discovered by Newton and gave a qualitative account of how the force of gravitation could explain the motion of the planets.

This lack of application of the calculus of fluxions to astronomy and mechanics is typical of the first generation of Newtonians. Despite the fact that in L'Hospital (1696), p. xiv, Newton's *Principia* was stated to be 'all about the calculus', and that Newton in the 'Account to the Commercium Epistolicum' (1715), p. 206, had affirmed that all the propositions of the *Principia* were demonstrated by his 'new analysis', it was not so easy to translate Newton's geometrical proofs into his fluxional calculus. The difficulties Leibniz and Varignon found in giving an analytic form to dynamics testify well to the fact that their effort was not trivial.

Little is known of Halley's activity as a teacher, while Bradley, Savilian Professor of Astronomy from 1721 to 1762, is described by Hans (1951), p. 48, and Turner (1986), p. 673, as lecturing regularly in the Ashmolean Museum. He was also lecturer on experimental philosophy in Christ Church. However, on his election in 1742 as Astronomer Royal, he had to leave Oxford. Bradley's lectures, as well as the preparatory sheets that he wrote for his own use, are still extant.[15] It appears that they were written well after the period considered in this chapter, i.e. in between 1758 and 1761. It will suffice here to say that Bradley's extant lectures cover schoolroom algebra up to logarithms and plane trigonometry. Another set of lectures includes matter theory (phenomena of cohesion and repulsion of particles), the first three laws of mechanics, motion on inclined planes, pendulums, central forces, elements of geometric optics, hydrostatics and pneumatics. Bradley copied a great part of his lectures from Rutherforth's *System of Natural Philosophy* (1748) and Emerson's *Principles of Mechanics* (1754), and he performed the experiments of Whiston and Hauksbee and derived the description of the necessary instruments from Desaguliers. From the preparatory notes of his lectures we gather that he intended to dedicate a few days to the calculus of fluxions. But, instead of referring his students to a treatise on fluxions, he chose a few examples taken from *Miscellanea Curiosa Mathematica*, a collection of short essays on mathematics edited by Francis Holliday (1745–53), in which one could find a translation of parts of Taylor's *Methodus Incrementorum* (1715). John Keill, James Bradley, Thomas Hornsby (their successor on the Savilian Chair of Astronomy), and John Whiteside (1679–1729), scientific lecturer at Christ Church and keeper of the Ashmolean Museum, gave lectures on natural philosophy (mechanics, hydrostatics, pneumatics, astronomy), but not on mathematics. As will be seen later, in the 1750s and 1760s the calculus of fluxions was studied in much more detail in Cambridge: it seems therefore that the grandiose projects of Gregory were never achieved.[16]

The position of Keill in Oxford resembles that of William Whiston (1667–1752) in Cambridge.[17] Whiston, in his *Praelectiones Astronomicae* (1707) and *Praelectiones Physico-Mathematicae* (1710), collected the lectures he gave as successor of Newton in the Lucasian Chair. In these works the importance of a mathematical treatment of planetary motions and geodesy (the shape of the Earth) is particularly stressed. However, Whiston did not attempt to employ the calculus and strictly adhered to the *Principia*: he reproduced the main propositions of Newton's masterpiece (and of the *Opticks*) and added his explanations. So the required mathematics is premised in the first three lectures and consists in the basic properties of the conic sections. In fact the *Praelectiones Physico-Mathematicae* are the first published extensive commentary of the *Principia*. It is not known if Whiston ever attempted a similar work with Newton's 'De quadratura', but this hypothesis seems very unlikely.

While we do not know anything about Cotes's and Whiston's teaching of the calculus, their activity as lecturers in experimental philosophy is very well documented. Since 1707 Whiston collaborated with Roger Cotes, Plumian Professor of Astronomy, in lecturing on experimental philosophy. Cotes's *Hydrostatical and Pneumatical Lectures* (1738) were published posthumously. It is known that Whiston, after having been deprived of his Chair because of his Arian heresy, moved to London, where he performed experiments with Francis Hauksbee (1687–1763) the younger. The lectures of Cotes, Hauksbee and Whiston are merely an example of the many courses on experimental philosophy which were given in that period. The number and success of lectures of this kind are easily explained by the fact that Newtonianism was more easily understandable, more attractive and spectacular when it took the form of an experimental course. Was, then, the teaching of the calculus of fluxions completely absent from Cambridge?

We can derive some information from two famous outlines of courses written in 1706 and in 1707.[18] The first one is the *Advice to a Young Student* (1730) by the theologian Daniel Waterland (1683–1740) of Magdalene. For our purpose it is interesting to note that Waterland included Well's '*Arithmetic, Geography* and *Astronomy*, Euclid's *Elements*, T. Newton's *Trigonometry*, de La Hire's *Conic Sections*, Whiston's *Praelectiones Astronomicae* and *Praelectiones Physico-Mathematicae*, Keill's *Introductio ad Veram Physicam*, Rohault's *Physics*, Newton's *Opticks* and Gregory's *Astronomiae Physicae & Geometricae Elementa*'. As further reading for those preparing for an M.A., Waterland added 'Ozanam's *Cursus Mathematicus*, Huygens's *Works*, Molineaux's *Dioptrica*, Harris's *Lexicon* and Newton's *Principia*'. As in the case of Gregory we do not have to take very seriously this full-time curriculum, but we have good reasons to think that at least *some* of the students found this guide useful. Indeed Waterland's *Advice* circulated in manuscript form was published in 1730 and was republished several times during the eighteenth century. But even though Waterland's *Advice* was the guide for the best students, we must still conclude that fluxions were almost absent from teaching at Cambridge.

However, the second guide we are considering here, Robert Greene's ΕΓΚΥΚΛΟΠΑΙΔΕΙΑ (1707), includes for the fourth year the study of 'fluxions and infinite series'. The authors to be consulted in this case are: Wallis, Newton, Raphson, Hayes, Ditton, Jones, Nieuwentijdt and L'Hospital. Robert Greene (1678-1736) was a fellow of Clare who distinguished himself as an anti-Cartesian, an anti-Newtonian and as a convinced supporter of a 'Greenian Philosophy'. Even though he was anything but a 'typical' figure his guide has to be taken into account. It is interesting, for instance, to find Nieuwentijdt and L'Hospital quoted again by a British author: Dutch and French mathematicians are very often quoted in early eighteenth-century Great Britain. Furthermore we find references to Hayes (1704), Ditton (1706) and Jones (1706).

Cotes died in 1716 and Whiston was expelled from Cambridge in 1710. Their successors, Robert Smith (1689-1768), Plumian Professor from 1716 to 1760, and Nicholas Saunderson (1682-1739), Lucasian Professor from 1711 to 1739, played an important role in establishing the study of mathematics in Cambridge.[19]

Nicholas Saunderson was blind from the age of twelve months as a consequence of smallpox. He was educated at the Attercliffe Academy, one of the most famous dissenting academies, where, it seems, he was not taught mathematics. But he found somebody who read him science books and, writing formulae and figures with pins on a wooden board, he acquired a deep knowledge of mathematics. He arrived at Cambridge in 1707 and began lecturing as a private tutor. Then in 1711, on the removal of Whiston, he was elected Lucasian Professor. A former student of his informs us that:

His Lecture, as soon as opened, was attended by many from several of the Colleges, and in some time was so crowded, that he could hardly divide the Day among all who were desirous of his Instructions. (Davies (1740), p. v)

We every Year heard the Theory of the Tydes, the *Phaenomena* of the Rainbow, the Motions of the whole Planetary System as upheld by Gravity, very well defended by such as had profited of his lectures. (Davies (1740), p. vi)

Students' copies of these lectures are still extant.[20] It appears that Saunderson, as was usual, lectured on a wide range of topics, i.e. mechanics, optics, hydrostatics, acoustics and astronomy. Here Saunderson did not add very much to Keill (1702) and (1718). Whiston (1710) and Worster (1722), the texts which he often quoted and even simplified in order to render his lectures accessible to innumerate students. However, Saunderson also lectured on mathematics, and probably because of his blindness he preferred analytical to geometrical methods.

His bias for analysis seems to have been particularly strong since it caused, as James Wilson (1690–1771) remembers, quarrels between Saunderson's and Benjamin Robins's students:

Amongst Mr. Robins's scholars, such as went afterwards to Cambridge, in order to qualify themselves for one of the learned professions, were wont, as in the custom of young men, frequently to enter into warm contests with the disciples of Mr. Prof. Saunderson, that gentleman using there a very different method of instruction. And indeed I have met with ingenious persons, who, though they allowed Euclid's Elements to be the perfectest book of the kind; yet did not think it the most proper introduction for the Generality of Students, at least when ranged into classes, the way of teaching principally followed in universities; but the contrary of this opinion appears to be true from the constant and very successful practice of the late famous Mr. Maclaurin, who, I observed with pleasure, always begun his academical courses with the Elements of Euclide. (Robins (1761), p. ix)

Saunderson lectured on algebra and fluxions. His *Elements of Algebra* and *Methods of Fluxions* were published posthumously in 1740 and 1756, respectively.

The *Method of Fluxions* (1756a) can be divided into three parts.[21] The first is an introduction to the calculus of fluxions. The second treats Cotes's integrals. The third (in Latin) is devoted to the analysis of some propositions of the *Principia*. The first part probably covers the contents of the course on fluxions given by Saunderson. The students would have been introduced to the calculus of fluxions at quite an advanced level. In addition to the topics covered by Hayes, Saunderson includes an interesting chapter on the attraction of spheroids which is an analytic treatment both of Newton's and Cotes's geometrical solutions. The analytical bias of Saunderson is particularly evident in the second part devoted to Cotes's 'Logometria' (1717). Although Cotes had given geometrical proofs, it appears that he used a table of eighteenth integrals which was published in *Harmonia Mensurarum* (1722).[22] Saunderson showed how Cotes's integrals could be derived one from the other and applied them to the solution of the propositions of the Scholium Generale of Cotes's 'Logometria'. The *Method of Fluxions* ends with a commentary of the *Principia* where he preserves Newton's geometrical methods. It is interesting that Saunderson does not confine his commentary to the first three sections of Book I, which would have been sufficient for an introduction to the mathematical treatment of gravity, but presents proofs of more advanced parts (for instance Prop. 66 and some of its corollaries). It seems that Saunderson was the first in Cambridge to lecture systematically on the calculus of fluxions.[23] The teaching of the calculus did not vanish after Saunderson. On the contrary, as we will see, in the 1740s, with the beginning of the Tripos Exam, exercises on fluxions became a routine for Cambridge students.

It seems likely that courses on the calculus of fluxions were given in Glasgow by Robert Simson (1687–1768), Professor of Mathematics from 1711 to 1761, and in Edinburgh by James Gregory (1666–1742?), who succeeded his brother David in 1692, and by Colin Maclaurin (1692–1746), who taught in Edinburgh from 1725 until 1746.[24]

Simson had as his pupils several mathematicians, most notably Colin Maclaurin (1698–1746), Matthew Stewart (1717–85), John Robison (1739–1805), James Williamson (d.1795) and William Trail (d.1831), who all held Chairs in the Scottish universities. He passionately advocated the use of geometrical methods. Trail remembers Simson's attitude towards the teaching of conic sections:

He had observed, in the first years of his study of Mathematics, that the treatises on Conic Sections, then in most general use and estimation, were entirely algebraical; and the great merit of the work, written in that stile by the Marquis De L'Hospital, contributed not a little to the popularity of this mode of treating geometrical subjects. It occurred therefore to Mr. Simson, that a Treatise on Conic Sections, written on the purer model of antiquity, might have some influence in correcting the prevailing false taste, of introducing algebraical calculation into those branches of geometry where it was not necessary, and where it supplanted a more elegant form of analysis and demonstration. (Trail (1812), p. 27)

Simson concerned himself with the restoration of Euclid's Porisms and with the edition of Euclid's *Elements* (1756): two tasks which occupied his entire life. However, we find him writing to Jurin (on 1 February 1732) about a series for the quadrature of the circle.[25] Furthermore, he left an incomplete tract, *De Limitibus Quantitatum et Rationum* (in Simson (1776), pp. 89–110), in which he tried to reformulate Newton's theory of prime and ultimate ratios within his preferred framework, i.e. he interpreted Newton's method as an abbreviated form of the indirect *ad absurdum* proofs of the method of exhaustion. Did Simson in his lectures on the *Principia* translate Newton's limit processes into an Archimedean framework? If so, in Glasgow the *Principia* would have been read as a Greek classic.

Even though Simson communicated his enthusiasm for geometry to Colin Maclaurin, it seems that his most famous pupil did not entirely agree with him. We know from James Wilson that Maclaurin always began his courses 'with the Elements of Euclide' (Robins (1761), p. ix). But we have reason to think that he soon left ancient geometry to teach his students something quite new. The plan of Maclaurin's *Treatise of Fluxions* (1742) reveals his intention to lead the reader from the first chapters concerned with the method of exhaustion to the second book which is completely analytical. He *might* have followed the same route in his courses. The doubtful nature of this conclusion must be stressed since the *Treatise of Fluxions* is too complex in structure and too advanced in content to be taken as representative of the teaching in Edinburgh.

It is interesting to see how John Playfair (1748–1819) in his biography (1788a) of Stewart, writing more than forty years after Maclaurin's death, contrasts Maclaurin's with Matthew Stewart's methodology:

Mr. Stewart's views made it necessary for him to attend the lectures in the University of Edinburgh in 1741; and that his mathematical studies might suffer no interruption, he was introduced by Dr. Simson to Mr. Maclaurin, who was then teaching, with so much success, both the geometry and the philosophy of Newton. Mr. Stewart attended his lectures, and made that proficiency which was to be expected from the abilities of such a pupil, directed by those of so great a master. But the modern analysis, even when so powerfully recommended, was not able to withdraw his attention from ancient geometry. (Playfair (1822a), IV, p. 5)

According to Playfair, Stewart would have noticed the difference between the study of ancient geometry in Glasgow and the 'new analysis' recommended in Edinburgh by Maclaurin. It would seem therefore that, while, in Glasgow, Simson taught Newton's philosophy to a number of young promising mathematicians approaching Newton's mathematics from a 'Euclidean' point of view, in Edinburgh Maclaurin promoted the calculus of fluxions starting, only as a background, from Euclid and Archimedes.[26]

Some further information on Maclaurin's teaching in Edinburgh can be gathered from the *Scots Magazine* for August 1741:

He [Maclaurin] gives every year three different colleges, and sometimes a fourth, upon such of the abstruse parts of the science as are not explained in the former three.

In the first, he begins with demonstrating the grounds of Vulgar and Decimal Arithmetic: Then proceeds to Euclid; and, after explaining the first six books, with the Plaine Trigonometry, and use of the tables of Logarithms, Sines, etc., he insists on Surveying, Fortification, and other practical parts; and concludes this college with the elements of Algebra. He gives Geographical lectures once in the fortnight to this class of students.

In the second college, he repeats the Algebra again from its principles, and advances farther in it; then proceeds to the theory and mensuration of Solids, the Spherical Trigonometry, the doctrine of the Sphere, Dialling and other practical parts. After this he gives the doctrine of the Conic Sections, with the theory of Gunnery, and concludes with the elements of Astronomy and Optics.

He begins the third college with Perspective; then treats more fully of the Astronomy and Optics. Afterwards he prelects on Sir Isaac Newton's *Principia*, and explains the direct and inverse method of Fluxions. At a separate hour he begins a college of Experimental Philosophy, about the middle of December, which continues thrice every week till the beginning of April; and at proper hours of the night describes the constellations, and shews the planets by telescopes of various kinds.[27]

Perhaps the universities were not the most appropriate places to find mathematics in the eighteenth century. The standard biography of the early eighteenth-century 'philomath' does not necessarily include a university education. In many cases mathematical education 'came to depend upon private enterprise' (Howson (1982), p. 59). Some schools specializing in mathematics were founded in the early eighteenth century: Sir Joseph Williamson's Free Mathematical School at Rochester (1701); Saunder's School at Rye (1708); Neale's mathematical School in Fleet Street, London (1715); Churcher's College at Petersfield (1722). However, it would be wrong to say that the teaching of David Gregory at Oxford, of Saunderson at Cambridge, of Simson at Glasgow and of Maclaurin at Edinburgh did not exert any influence on the development of eighteenth-century British mathematics. In Oxford mathematics soon faded after Gregory's death. But Saunderson and Smith established the study of mathematics in such a way that Cambridge began to produce mathematicians in the second half of the century (Waring, Cavendish, Atwood, Brinkley and Maskelyne, for instance, were Cambridge students). Simson in Glasgow educated several of the future professors of mathematics of the Scottish universities, Colin Maclaurin and Matthew Stewart being two of his pupils. Finally, the activities of Maclaurin in Edinburgh as Professor of Mathematics gained for mathematics an important place in the Scottish Enlightenment.[28] As we will see in chapter 7, Edinburgh became one of the most influential centres of reform of the British calculus. These early attempts to diffuse the calculus in the universities, which began more or less with Newton's death, laid the foundations of important trends in British mathematical education.

2

DEVELOPMENTS IN THE CALCULUS
OF FLUXIONS (1714–33)

DESPITE ITS minor role in the diffusion of Newtonian science, we might
expect the calculus of fluxions to have been the main subject of research
in early eighteenth-century British mathematics. Scientists of outstanding
ability were active in Great Britain. Some of them, such as Cotes, Taylor,
Maclaurin and Stirling, were good mathematicians who were able to
master every aspect of Newton's work on fluxions. But British mathematicians devoted more attention to other branches of Newton's
mathematics: i.e. the geometry of higher order curves and the method of
series. The reason why the calculus of fluxions was not considered a
fruitful area of research is that it appeared to have been developed by
Newton to the highest level of perfection. For instance, in 1721 Colin
Maclaurin wrote:

The Quadratures [i.e. Newton (1704c)] brought to such generall [sic] theorems
that little further seems left to be done in that vast feild [sic] of Invention.
(Maclaurin (1982), p. 13)

The direct method allowed one to find the fluxion of all the known fluents,
whereas the inverse method required term by term integration of power
series. One of the problems left open was to speed up the convergence of
series; a problem which could be treated by finding appropriate
transformations involving finite differences. Newton's 'Methodus
differentialis' (1711c) appeared more incomplete than 'De quadratura'
(1704c). Also, Newton's 'Enumeratio' (1704b) was more problematic in
the classification of cubics. A great interest in the geometry of higher order
curves is characteristic throughout the eighteenth century in Great
Britain. This subject deserves more attention, but does not fit into the
scheme of the present work. The two British mathematicians who, in the
period taken into consideration in this chapter, committed themselves to
the project of extending the calculus of fluxions were Taylor and Cotes.

They were both motivated by the mechanics of Newton's *Principia* (1687). Probably for this reason they realized that integration was an open field of research.

2.1 Methods of integration

Almost every aspect of the exact sciences in the eighteenth century can be traced back to Newton's *Principia*. The attention of natural philosophers was immediately attracted by the novelty of the mathematical tools, the universality of the laws of mechanics and the power of the cosmology of the *Principia*. These interests originated the well-known controversies about the nature of the calculus, the conservation of dead or live forces, the causes of planetary motions. However, it was perhaps more stimulating that in the *Principia* Newton had created and left unanswered some very specific problems. It was clear that further explanations were required on the motions in resisting mediums, the ebb and flow of the tides and the irregularities of the Moon's orbit: three aspects of Newton's theory where there was a great discrepancy between theory and observation. In order to find a satisfactory solution to these problems, both the mathematics and the mechanics of the *Principia* had to be completely transformed: to a great extent the theory of partial differential equations, the calculus of variations and continuum mechanics have their origin in this process.

If it is legitimate to see the progress of eighteenth-century mathematics and mechanics as a criticism and development of the ideas and the problems of the *Principia*, then Roger Cotes (1682–1716) occupies a privileged position amongst eighteenth-century mathematicians.[1] He was involved from 1709 to 1713 in the preparation of the second edition of the *Principia*, an edition which as a result of his contribution differs appreciably from the first. In 1707 Cotes was named first Plumian Professor at Cambridge. As we have already seen, he started lecturing with Whiston on experimental philosophy.[2] He also concerned himself with the construction of an astronomical observatory at Trinity College, but his premature death did not allow him to see this project through to its completion. This work was continued by Cotes's cousin, Robert Smith, who was his successor in the Plumian Chair and the editor of his mathematical works.

Cotes's first published work appeared as a long article in the *Philosophical Transactions* for the year 1714 with the title 'Logometria' (1717). This was reprinted as the first part of *Harmonia Mensurarum* (1722). Both the methods and the contents of 'Logometria' reveal the influence of the years of work on Newton's *Principia* and a deep knowledge of Newton's 'De quadratura' (1704c). In fact many of the problems solved have their origin in Newton's work. The 'Logometria' consists of six propositions with

various *scholia* and a *Scholium Generale* which concludes the work. In the
six propositions Cotes presents a series of results concerning logarithms.
Here we find a calculation of the natural base for a system of logarithms,
the result being 2.718281828459. After several theorems on systems of
logarithms he considers the properties of the hyperbola, the logarithmic
curve and the equiangular spiral in terms of logarithms and applies these
results to the study of the vertical ascent and descent in resisting mediums,
to the determination of the density of the atmosphere as a function of
altitude, and calculates the change in longitude as a function of the change
in latitude along a loxodrome.[3] In the *Scholium Generale* Cotes gives
geometric solutions to a series of problems on the rectification of curves,
areas, surfaces and volumes. He also considers mechanical problems such
as the gravitational attraction of ellipsoids and the oscillations of cycloidal
pendulums in resisting mediums.

Even though these results are given in geometric form, they were
obtained using a table of integrals which was published in the second part
of *Harmonia Mensurarum* (1722).[4] In his tables Cotes uses a notation which
expresses the 'harmony' between measures of angles and measures of
logarithms. Cotes uses:

$$R \left| \frac{R+T}{S} \right. \text{ for } R \times \ln\left(\frac{R+T}{S}\right)$$

when R^2 is positive, while if R^2 is negative, i.e. if R is imaginary, it expresses
$|iR\mathrm{arctan}(T)|$. Cotes integrates 'forms' such as:

$$\frac{dz^{\mu a + \frac{1}{2}a - 1}}{\sqrt{(e + fz^a)}} \dot{z},$$

where d, e, f and a are constants, μ can take integer values, and z is the
'fluent quantity', or, as we would put it nowadays, z is variable. These
forms are generalizations of the integrals ('fluents') tabulated after Prop.
10 of Newton's 'De quadratura'. But whereas Newton had reduced his
fluents to the quadrature of conic sections, Cotes performed his integrations
in terms of logarithms and trigonometric functions. Furthermore his
notation allowed him to tabulate in a systematic way his integrals, or as
he called them 'forms of fluents'. It is difficult to say if Cotes was aware of
$iy = \ln (\cos y + i\sin y)$. He was certainly able to understand that his tables
afforded a means of studying the relationships between circular and
logarithmic functions. The most important result in *Harmonia Mensurarum*
(1722) was the so-called Cotes factorization theorem, which allows one to
find the factors of $a^n \pm x^n$, $n \in N$.[5] The factorization theorem was employed
by Robert Smith to extend Cotes's integration formulae to 'forms' such as:

$$\frac{dz^{\mu a + \frac{3}{4}a - 1}}{e + fz^a + gz^{2a} + hz^{3a}}\dot{z}, \ \mu \in \mathbb{Z};$$

d, e, f, g, h and a are constants, z is a variable.[6]

Smith added ninety-four new integrals to those obtained by Cotes and inserted them in the fourth part of *Harmonia Mensurarum* (1722). He also added a series of incomplete works on estimating errors, finite differences and summations written by Cotes presumably just before his death. de Moivre (1667–1754) was working along the same lines.[7] In (1708) he stated a formula related to the well-known de Moivre formula. In (1730) he applied it to derive Cotes's factorization theorem.

Cotes's work is extremely important and, in a way, exceptional for eighteenth-century British mathematics. In solving mechanical or geometrical problems, Cotes proceeds by geometric arguments in order to reduce the problem to a quadrature. This was typical of Newton who in the *Principia* began his propositions 'granting the quadrature of figures'. Only after this reduction is the calculus employed in order to measure that particular area, arclength, etc. which answers the question. Cotes's integrals are interesting because they are in finite form and because they allow an understanding of the relationships of circular and exponential functions; however, both these aspects were not really appreciated by the fluxionists, who, during the eighteenth century, considered Cotes's tables of integrals as merely an expedient method of integration.

2.2 The *Methodus Differentialis*

Another Cambridge man, Brook Taylor (1685–1731), is to be included amongst the most creative British mathematicians of the early eighteenth century. Taylor entered St John's College in 1701 and graduated L.L.B. in 1709 and L.L.D. in 1714. He was secretary of the Royal Society from 1714 to 1718. His major work, the *Methodus Incrementorum Directa & Inversa* (1715), was published in London in 1715 and influenced many mathematicians in the first half of the eighteenth century.[8]

Taylor's *Methodus* is divided into two parts. The first is devoted to Taylor's theorem, and Newton's method of integration by series and summation formulas. The second deals with interpolation and with the solution of mechanical problems, such as finding the centres of oscillation (see also Taylor (1714a)) and percussion, the vibrating string problem (see also Taylor (1714b)), and the calculation of the density of the atmosphere.

Taylor employed x, $\underset{.}{x}$, $\underset{..}{x}$, etc. to designate the finite differences Δx, $\Delta^2 x$, $\Delta^3 x$, etc. and $\dot{x}$, $\ddot{x}$, $\dddot{x}$, etc. to designate the fluxions of x. It is interesting that he used positive and negative integers as superscripts and subscripts for higher order differences and fluxions. He designated the successive values of x by $\overset{\cdot}{x}$, $\overset{\cdot\cdot}{x}$, x, $\underset{\cdot}{x}$, $\underset{\cdot\cdot}{x}$, etc. Taylor's notation is a proof that it was quite possible to modify the fluxional notation in order to approach the calculus of fluxions as a calculus of operations upon successions of variables, a characteristic usually ascribed to the differential notation.[9] The notation employed in the *Methodus* also expressed Taylor's belief that the calculus of fluxions was just a particular case of a more general theory, the calculus of finite differences, envisaged by Newton in his 'Methodus differentialis' (1711c).

Taylor described his (1715) in the *Philosophical Transactions* as follows:

When I apply'd my self to consider throughly the Nature of the Method of Fluxions, which has justly been the Occasion of so much Glory to its great Inventor Sir Isaac Newton our most worthy President, I fell by degrees into the Method of Increments, which I have endeavour'd to explain in this Treatise. For it being the Foundation of the Method of Fluxions that the Fluxions of Quantities are proportional to the nascent Increments of those Quantities: in order to understand that Method throughly, I found it necessary to consider well the Properties of Increments in general. And from those Properties I saw it would be easy to draw a perfect Knowledge of the Method of Fluxions: for if in any case the Increments are supposed to vanish and to become equal to nothing, their Proportions become immediately the same with the Proportions of the Fluxions. (Taylor (1717), pp. 339–40)

An example of this passage from a general proposition on increments to a proposition on fluxions is given by the proof of Taylor's theorem. In the *Methodus* Taylor, starting with a generalization of a formula of interpolation given in the *Principia*, obtained his famous theorem by simply passing from finite differences to fluxions. Even though similar results can be found in James Gregory, Newton, Leibniz and Johann I Bernoulli, it seems that he was the first to appreciate the importance of this theorem. It is worthwhile considering Taylor's proof, because it is characteristic of the elegance and heuristic power of his calculus. Proposition VII reads as follows:[10]

Let z and x be two variable quantities, of which z is increased uniformly by the given increments $\underset{\cdot}{z}$, and let $n\underset{\cdot}{z} = v$, $v - \underset{\cdot}{z} = \overset{\cdot}{v}$, $\overset{\cdot}{v} - \underset{\cdot}{z} = \overset{\cdot\cdot}{v}$, and so on. Then I say that in the time that z increases to $z + v$, x will likewise increase to

$$x + \underset{\cdot}{x}\frac{v}{1 \cdot \underset{\cdot}{z}} + \underset{\cdot\cdot}{x}\frac{v\overset{\cdot}{v}}{1 \cdot 2 \cdot \underset{\cdot}{z}^2} + \underset{\cdot\cdot\cdot}{x}\frac{v\overset{\cdot}{v}\overset{\cdot\cdot}{v}}{1 \cdot 2 \cdot 3 \cdot \underset{\cdot}{z}^3} + \ldots.$$

(Taylor (1715), p. 21; translation by Feigenbaum (1985), p. 40)

In Corollary 1 is stated the form of Proposition VII for a 'decrease' of z. So, if z decreases to $z - v$, x will decrease to

$$x - \dot{x}\,\frac{v}{1 \cdot \dot{z}} + \ddot{x}\,\frac{v\dot{v}}{1 \cdot 2 \cdot \dot{z}^2} - \dddot{x}\,\frac{v\dot{v}\ddot{v}}{1 \cdot 2 \cdot 3\dot{z}^3} + \dots .$$

Then in Corollary 2 Taylor obtains his famous result:[11]

If, in place of the evanescent increments, the fluxions proportional to them are written, and if $\dddot{v}$, $\ddot{v}$, $\dot{v}$, … are now made equal, then in the time that z, flowing uniformly becomes $z + v$, x will become

$$x + \dot{x}\,\frac{v}{1 \cdot \dot{z}} + \ddot{x}\,\frac{v^2}{1 \cdot 2 \cdot \dot{z}^2} + \dddot{x}\,\frac{v^3}{1 \cdot 2 \cdot 3 \cdot \dot{z}^3} + \dots .$$

(Taylor (1715), p. 23; translation by Feigenbaum (1985), p. 42)

This last passage has caused some historians much perplexity. However, for Taylor's contemporaries it appeared perfectly understandable and intuitively sound. Indeed the calculus seen as a limiting case of a theory of finite increments seemed to many mathematicians less mysterious than the calculus of moments or fluxions.

A similar interest in Newton's theorems on finite differences motivated James Stirling (1692–1770) in his research. Stirling belonged to an aristocratic Scottish family connected with the Jacobites. He moved to Oxford in 1711 where he acquired great fame as a mathematician, as is clearly shown by the number of Oxford men listed as subscribers of his first work *Lineae Tertii Ordinis Neutonianae* (1717). It is possible that, because of his political views, he had to leave Oxford in 1716 without graduating.[12] Later we find him in Italy where he might have had contact with the University of Padua.[13] His stay in Italy ended in 1724. Stirling then moved to London where he accepted the offer to succeed Benjamin Worster as a teacher of mathematics in Watts' Academy, initially a school for clerks and accountants. In 1730 he published his masterpiece the *Methodus Differentialis* (1730). Stirling's scientific activity finished in the 1730s. From 1735 to his death he worked as an administrator of the lead mines at Leadhills, Lanarkshire; and so he spent the rest of his life in isolation without pursuing his scientific activities.[14]

Stirling's *Methodus Differentialis* (1730) greatly extends Newton (1711c). It consists of an introduction and is in two parts; the aim of the first part is to speed up the convergence of series, while the second is devoted to interpolation.

In the introduction Stirling presents the notation for infinite series:

$$S = T' + T'' + T''' + \dots ,$$

and declares his intention of dealing with series of the form:

$$(*)\ T = A + Bz + Cz(z-1) + Dz(z-1)(z-2) + \dots$$

and

$$(**)\ T = A + B/z + C/(z(z+1)) + D/(z(z+1)(z+2)) + \dots.$$

In fact he states:[15]

Of course, the former formula should be used when z is a small quantity, while the latter when z is large. And these series which are composed of factors in arithmetical progression are much more useful than the ordinary ones which consist of ascending or descending powers of the unknown quantity. (Stirling (1730), p. 6)

So we can see that Stirling is trying to study series of the form $(*)$, $(**)$, because in many cases the power series which were usually employed converge too slowly. Therefore he needs a method of facilitating the conversion of powers into factorials and vice versa; he obtains this method in two ways, which written in modern notation are (see Tweedie (1822), p. 31):

$$z^n = \Gamma_2^{n-1}z + \Gamma_3^{n-2}z(z-1) + \dots + \Gamma_{n+1}^0 z(z-1)\dots(z-n+1),$$

$$z^{-n} = \sum_{r=n-1}^{\infty} C_r^{r-n+1}/(z(z+1)\dots(z+r)),$$

where Γ_n^s and C_n^r represent the Stirling's numbers.[16] Stirling does not use such a general notation, but gives the examples for the lower powers, e.g. he writes (Stirling (1730), p. 8):

$$z^3 = z + 3z(z-1) + z(z-1)(z-2)$$
$$z^4 = z + 7z(z-1) + 6z(z-1)(z-2) + z(z-1)(z-2)(z-3)$$
$$z^5 = z + 15z(z-1) + 25z(z-1)(z-2) + 10z(z-1)(z-2)(z-3)$$
$$+ z(z-1)(z-2)(z-3)(z-4).[17]$$

As an example of factorial representation of series Stirling gives (p. 12):

$$\frac{1}{z^2 + nz} = \frac{1}{z(z+1)} + \frac{1-n}{z(z+1)(z+2)} + \frac{2 - 3n + n^2}{z(z+1)(z+2)(z+3)} + \dots.$$

Other important results on speeding up the convergence of series are to be found in part I *de Summatione Serierum*, pp. 15–84 (see Krieger (1968)).

In the second part *de Interpolatione Serierum*, pp. 85–153, after some introductory pages devoted to Newton's interpolation formulae and to Taylor's theorem, Stirling deals with two important results, nowadays expressed as:

$$\Gamma(\tfrac{1}{2}) = \sqrt{\pi},$$

$$n! = n^n e^{-n} \sqrt{(2\pi n)}\, e^{\theta_n/(12n)}, \quad 0 < \theta_n < 1.$$

The former arises from the problem of interpolating the factorials 1, 1, 2, 6, 24, 120, 720, etc., the law of succession being $T_{z+1} = zT_z$, with $T_1 = 1$. Stirling considers the logarithms of the T_j so as to facilitate the accuracy of the interpolation:[18]

I propose to find the term which is in the middle between the first two, 1 and 1. And since the logarithms of the initial terms have slowly converging differences, I will first determine the term which is in beween two terms distant enough from the beginning, for instance between the eleventh 3628800 and twelfth 39916800; and from this I will go back to the required term. (Stirling (1730), p. 110)

Once a value for $\log(T_{11+1/2}) = 7.07552590569$ has been found by interpolation, Stirling divides $T_{11+1/2}$ by 10·5, 9·5, ..., 1·5 to obtain $T_{1+1/2} = 0.8862269251$, and its double $T_{1/2} = 1.7724538502$. Finally, he observes:[19]

Therefore the term between 1 and 1 is 0·8862269251; whose square is 0·7853...etc., which is of course the area of a circle whose diameter is equal to 1. And twice that term, 1·7724538502 [...] is equal to the square root of the number 3·1415926...etc., which denotes the circumference of the circle whose diameter is equal to 1. (Stirling (1730), pp. 112–13)

The second result is on pp. 135–7. Here Stirling's aim is to calculate

$$\log(x+n) + \log(x+3n) + \log(x+5n) + \ldots + \log(z-n).$$

He expresses this sum as the difference of two series:

$$\frac{z\log(z)}{2n} - \frac{az}{2n} - \frac{an}{12z} + \frac{7an^3}{360z^3} - \frac{31an^5}{1260z^5} + \ldots$$

and

$$\frac{x\log(x)}{2n} - \frac{ax}{2n} - \frac{an}{12x} + \frac{7an^3}{360x^3} - \frac{31an^5}{1260x^5} + \ldots,$$

where $a = 1/\ln 10$. In the following Example II Stirling applies this result to the calculation of $\log(x!)$ and obtains (see Tweedie (1922), p. 43):

$$\log(x!) = \tfrac{1}{2}\log(2\pi) + (x + \tfrac{1}{2})\log(x + \tfrac{1}{2}) - (x + \tfrac{1}{2})$$

$$-\frac{1}{2 \cdot 12 \cdot (x + \tfrac{1}{2})} + \frac{7}{8 \cdot 360(x + \tfrac{1}{2})^3} - \ldots$$

This series was to be reformulated in many other forms, and has caused quite a lot of problems as to its convergence (e.g. see Bayes (1764)). In fact the series is divergent, but it can be used to approximate $\log(x!)$ if x is large. Typically, Stirling does not perceive any difficulty and develops his proof by purely algebraical manipulations.[20]

2.3 Geometry

The first systematic work devoted to Newton's classification of cubics is Stirling's *Lineae Tertii Ordinis Neutonianae* (1717).[21] It deserves attention for several reasons. In it Stirling applied the calculus to the study of the seventy-two species of cubic curves classified by Newton and added four new ones (which did not appear in print in Newton (1704b), even though Newton had already obtained them). The work was completed by an appendix on three topics which by 1720 were routine exercises: i.e. the study of the brachistocrone, the catenaria and orthogonal trajectories to a family of hyperbolas.[22] The introductory part on fluxions (pp. 6–40) was a useful comment of Newton's 'De analysi' (1711b). Stirling devoted particular attention to Newton's methods of finding power series representations of fluents. He then proceeded (pp. 41–83) to explain how the calculus of fluxions could be applied to the study of curves: finding zeros, asymptotes, cusps, points of contact, etc. After this introductory material Stirling moved on to consider Newton's 'Enumeratio linearum tertii ordinis'. This is interesting because compared to Newton, who had given a geometric treatment of cubics, Stirling employed the calculus. Therefore, Stirling not only commented on the 'De analysi' (1711b) but also explained how the analytical methods of the calculus of fluxions could be successfully applied where Newton had preferred geometry.

Three years after Stirling's first work, another Scottish mathematician, Colin Maclaurin (1698–1746), entered the scene of British mathematics with a work devoted to Newton's geometrical work, the *Geometria Organica* (Maclaurin (1720c)).[23] In contrast with Stirling, Maclaurin did not use the calculus of fluxions. He had learnt mathematics at the University of Glasgow, which he had entered in 1709 at the age of eleven. Robert Simson, who in 1711 had been elected Professor of Mathematics in Glasgow, took care of young Colin's education and communicated to him a great esteem for Greek geometry. In 1717 Maclaurin was chosen for the Chair of Mathematics in Marischal College in Aberdeen. However, he did not fulfil his duties since in 1719 we find him in London and from 1722 to 1724 in France. In London Maclaurin met Newton who, as President of the Royal Society, approved the publication of the *Geometria Organica*. In France, where he went as a tutor to the son of Lord Polwarth, he was awarded a prize from the Académie des Sciences for a dissertation on the percussion of bodies.[24] On his return to Scotland, he was elected, through the recommendation of Newton, Professor of Mathematics in the University of Edinburgh, a post he held up to his death in 1746. In Edinburgh Maclaurin wrote *A Treatise of Fluxions* (1742) in which he collected together a great deal of his research. He is also the author of *A Treatise of Algebra* (1748b), a commentary to Newton's *Arithmetica Universalis*, and *An Account of Sir Isaac Newton's Philosophical Discoveries* (1748a), one of the most popular introductions to Newton's natural philosophy.

In the *Geometria Organica* (1720c) Maclaurin extended Newton's organic description of cubics, in which curves were described by the motions of given angles, to curves of a higher order. Even though Maclaurin obtained his results without the use of the calculus, or any geometric procedure equivalent to the techniques of the calculus, on pp. 120–35 he gave some theorems on central forces 'in order to show the use of curve lines in Natural philosophy'. Here Maclaurin used dotted letters to represent 'momenta', i.e. infinitesimal quantities generated by motion.[25] These pages, derived from Keill (1710), are completely separate from the rest of the work since no attempt is made to link the geometry of Newton's 'Enumeratio' (1704b) with the analysis of 'De quadratura' (1704c). As we will see later, one of the main objectives of Maclaurin's research in the 1730s was to systematize in a comprehensive theory the various aspects of Newton's mathematical work.

The geometrical bias of the young Maclaurin was not an exception in early eighteenth-century British mathematics. Newton, Halley and Simson were deeply concerned with the restoration of the works of Greek geometers: in particular, the restoration of Euclid's *Porisms* became a dominant programme, which they inherited from their sixteenth-century predecessors like Commandinó. They were motivated by the genuine belief that the geometrical analysis of the ancients was superior to the modern techniques of the calculus. The ancients were thought to have concealed their *resolutio* and to have presented only the synthetic demonstrations. As Newton wrote:

Indeed their method is more elegant by far than the Cartesian one. For he achieved the results by an algebraical calculus which, when transposed into words (following the practice of the Ancients in their writings), would prove to be so tedious and entangled as to provoke nausea, nor might it be understood. But they accomplished it by certain simple propositions, judging that nothing written in a different style was worthy to be read, and in consequence concealing the analysis by which they found their constructions. (Newton (1967–81), IV, p. 277)

The myth of the power of Greek geometry was part of a more general attitude towards ancient science. It is known that Newton was convinced that the ancients had discovered the general laws of the motions of the planets.[26] This programme, rather than the Newtonian mechanics and astronomy, motivated Robert Simson, the young Colin Maclaurin, and later Matthew Stewart.

3

THE CONTROVERSY ON THE
FOUNDATIONS OF THE CALCULUS
(1734–42)

THE PROBLEM of foundations did not exist in the eighteenth century as we understand it nowadays. Mathematicians were more occupied with defining the 'principles' of the calculus. They were concerned with the ontological status of the objects of the calculus and with the correctness of the methods of the calculus according to the standards of Aristotelian logic. In 1734 these issues were raised by Berkeley in *The Analyst* (1734). His criticisms were particularly devastating in Britain since the terminology of the fluxionists was very loose, being a mixture of Newtonian and Leibnizian ideas. A controversy between Berkeley and some defenders of Newton occupied the years 1734–5. Later the debate divided the fluxionists themselves. These years of debate were of great importance since the British were compelled to reread Newton in order to remove from his calculus the flaws indicated by Berkeley. The great champion of this process of reinterpretation was Maclaurin. After his *Treatise* (1742) the calculus of fluxions solidified into a theory based on an 'axiomatization' of the basic properties of motion and velocity in which geometric limit processes were prominent.

3.1 Berkeley's criticisms of the calculus

The most famous aspect of the history of eighteenth-century British calculus is undoubtedly the controversy originated by *The Analyst* (1734). As is well known, George Berkeley (1685–1753) criticized in this short pamphlet the foundations of the fluxional and the differential calculus. The contents of *The Analyst* have already been analysed in detail in Grattan-Guinness (1969): here it will be sufficient to summarize the main steps of Berkeley's criticism of the calculus.

Berkeley never denied that the calculus could successfully solve complex geometrical and mechanical problems, but he considered unsatisfactory the definitions of the terms and the justifications for the methods of proof. In his opinion, the calculus was far from being as exact and certain as the other branches of mathematics: if correct results were achieved, it was because of a fortunate compensation of errors. Berkeley did not therefore direct his criticism at particular aspects of the calculus: his objective was to denounce the lack of rigour in the foundations of the new analysis of Leibniz and Newton.[1]

Berkeley's criticisms can be divided into two types: ontological and logical. In the ontological criticisms Berkeley argued that the objects to which the calculus refers do not exist. These criticisms were important for the fluxionists: it was difficult for them to attribute a scientific character to a theory in which 'meaningless' symbols were employed. British mathematicians had an empiricist philosophical background. A theory, in order to be accepted as scientific, had to avoid reference to fictitious, hypothetical entities, such as Leibnizian vortices. Newton's cosmology was maintained to be superior to Leibniz's because Newton would have carefully banished terms devoid of empirical meaning. Even though nobody tried to develop an empiricist methodology of mathematics, it was somehow implied that mathematics too had to possess a certain empirical foundation.[2] In particular, the calculus of fluxions was thought to be a theory dealing with continuously varying magnitudes, and, as we know, the study of continuous magnitudes could be carried on by using infinitesimals (our (1) and (2), see p. 6) or rates of increase (our (3) and (4)). Berkeley directed his attention to both.

Berkeley first of all equated (1) (differentials) and (2) (moments); he wrote:

The Points or mere Limits of nascent Lines are undoubtedly equal, as having no more magnitude one than another, a Limit as such being no Quantity. If by a Momentum you mean more than the very initial Limit, it must be either a finite Quantity or an Infinitesimal. But all finite Quantities are expressly excluded from the Notion of a Momentum. Therefore the Momentum must be an Infinitesimal. (Berkeley (1734), p. 18)

The notion of infinitesimals is not empirically founded according to Berkeley's sense-data-based theory of knowledge: the infinitesimals are, by definition, beyond the *minimum sensibile*; they are outside the scope of our understanding of existence. From this point of view, a calculus of infinitesimals dealt with non-existing entities and therefore was devoid of any scientific character. The fluxionists had the alternative of accepting as terms of reference in their calculus only finite rates of change (our (3)) or velocities (our (4)). But also in these cases, Berkeley thought, the 'analysts' went beyond the limits of our knowledge: we can perceive and measure finite changes of position in space and finite intervals of time. The term 'velocity' is empirically founded if used to mean the ratio of a finite space and a finite time, but we cannot have any appreciation of an instantaneous velocity:

A Point may be the limit of a Line: a Line may be the limit of a Surface: a Moment may terminate Time. But how can we conceive a Velocity by the help of such Limits? It necessarily implies both Time and Space, and cannot be conceived without them. And if the Velocities of nascent and evanescent Quantities, i.e. abstracted from Time and Space, may not be comprehended, how can we comprehend and demonstrate their Proportions? Or consider their *rationes primae* and *ultimae*? For, to consider the Proportion or *Ratio* of Things implies that such Things have Magnitude: that such their Magnitudes may be measured, and their Relations to each other known. (Berkeley (1734), pp. 50–1)

It is not possible to attribute existence to what is beyond the scope of our perception, and fluxions, as long as they are defined as ultimate ratios of vanishing quantities, are by definition something which our senses cannot reach. Furthermore, according to Berkeley's operational view of physical magnitudes, the concepts of space and time have a meaning only if they are defined as results of measurements: this was one of the points of Berkeley's criticism of Newtonian absolute time and space. Therefore the notion of velocity could be accepted only as the result derived from measuring a finite space and a finite time.

Berkeley did not concern himself only with the ontological question of the referents of the terms 'fluxion', 'moment' and 'differential'; he also turned his attention to the deductive techniques of the calculus. His logical criticism, both of the calculus of limits and of the calculus of infinitesimals, was that the conclusions were obtained by taking as given a hypothesis and its negation. As was well known from scholastic logic anything could follow from a contradiction. He wrote:

If, with a View to demonstrate any Proposition, a certain Point is supposed, by virtue of which certain other Points are attained; and such supposed Point be it self afterwards destroyed or rejected by a contrary Supposition; in that case, all the other Points, attained thereby and consequent thereupon, must also be destroyed and rejected, so as from thence forward to be no more supposed or applied in the Demonstration. (Berkeley (1734), pp. 19–20)

Berkeley tried to show that the 'analysts' were guilty of this fallacy. In the calculus of infinitesimals the ambiguous nature of the differentials served to hide the fact that the demonstrations were based from the outset on the supposition that $dx \neq 0$, a supposition which was then denied at a crucial point. In fact, when the principle of cancellation of higher order differentials

$(d^n x + d^{n+1} x = d^n x)$ was employed, differentials were treated as zeros and the quantities multiplied by them were eliminated from the calculation. Similarly, in the calculus of limits there was a 'shifting' of a supposition: once again a quantity introduced as different from zero was equated to zero in the middle of the calculation.

Berkeley reached a very high standard of accuracy in his logical analysis, and his proposal to consider the calculus as grounded on a compensation of errors exerted a certain influence in the history of the eighteenth-century calculus.[3] The importance of *The Analyst* for the development of the British calculus can be evaluated by comparing the careless approach to foundations of the fluxionists before 1734 with the immediate answers to Berkeley's criticisms.

3.2 The definitions of the basic terms of the calculus in the works of the early fluxionists

The problem of foundations was never seriously treated before Berkeley. An example of the careless use of definitions which characterizes the works of the early fluxionists can be seen in Stone (1730), possibly a source for Berkeley. Stone wrote in the Preface to his translation of L'Hospital's *Analyse des Infiniment Petits* (1696):

[almost all the Foreigners] represent the first Increment, or Differential (as they call it) by the letter d, the second by dd, the third by ddd, &c.; the fluents, or Flowing Quantities, being called Integrals. But since this method in the Practice thereof, does not differ from that of Fluxions, and an Increment or Differential may be taken for a Fluxion; out of regard to Sir Isaac Newton, who invented the same before the year 1669, I have altered the Notation of our Author, and instead of d, dd, d^3, &c. put his Notation, viz. $\dot{x}$, $\ddot{x}$, $\dddot{x}$, &c. or some other of the last Letters of the Alphabet, printed thus, and called the infinitely small Increment, or Differential of a Magnitude, the Fluxion of it. (Stone (1730), p. xviii)

Indeed, as Newton had warned in the 'Account' to the *Commercium Epistolicum* (1715), $\dot{x}$, a finite velocity, cannot be taken for dx, an infinitesimal. However, as De Morgan pointed out in his (1852), equating fluxions and differentials was a common practice in this early period; and this created a great confusion in the terminology of early fluxionists.

For instance, Joseph Raphson in his *Mathematical Dictionary* (1702) begins the article 'Fluxions' as follows:

Fluxions, in *Geometry* is a new improvement of it upon the doctrine of *Indivisibles* and *Arithmetic* of *Infinites*. (Raphson (1702), page unnumbered)

Newton had tried to distinguish between a calculus of moments and a calculus of limits: in particular 'De quadratura' (1704c) was an attempt to base the calculus solely upon limits. But generally Newton's effort was not understood.[4]

John Harris, who inserted an English translation of Newton's 'De quadratura' in the second volume of the *Lexicon Technicum* ((1704, 1710), II, 'Quadrature of curves'), did not feel the need to change the substance of the article 'Fluxions' published in the first volume (1704) where we read:

by the *Doctrine of Fluxions*, we are to understand the Arithmetick of the *Infinitely small* Increments or Decrements of *Indeterminate* or *variable Quantities*, or as some call them the *Moments* or *Infinitely small Differences* of such variable Quantities. These Infinitely small Increments or Decrements, our Incomparable Mr. Isaac Newton, calls very properly by the Name of Fluxions. (Harris (1704, 1710), I, 'Fluxions')

In the second volume Harris continued the article 'Fluxions' with a 'general Method of finding the Fluxions of all Powers and Roots, I had from the Honourable Fr. Robartes, Esquire'. Harris wrote:

If a Quantity gradually increases or decreases, its immediate Increment or Decrement is called its *Fluxion*. Or the Fluxion of a Quantity is its Increase or Decrease indefinitely small. (Harris (1704, 1710), II, 'Fluxions')

The conception of continuity adopted by the early fluxionists was in the great majority of cases that of Newton's 'De analysi': i.e. they based the calculus on infinitesimals. For instance, Jones wrote in *Synopsis Palmariorum Matheseos* (1706):

all *Curved Lines* may be considered as composed of an Infinite Number of Infinitely little right Lines: And any one of them Produced, only Touches the Curve, therefore is called the *Tangent* of that Point of the *Curve*. (Jones (1706), p. 226)

The use of infinitesimals and the lack of a clear distinction between fluxions and differentials was so common that it would be possible to add many other quotations.[5] Thinking about the calculus in terms of infinitesimals was so natural for early Newtonians that John Keill, as we have already seen one of the most vehement defenders of Newton's priority over Leibniz, could maintain the physical existence of 'fluxions of fluxions, or infinitesimals of infinitesimals' (see chapter 1, section 1.3).

Taylor and Stirling came closer to 'De quadratura' (1704c). Even though their observations on the nature of the calculus are not completely coherent, they sometimes hinted that the calculus of fluxions was to be conceived as a limiting case of the calculus of finite differences. In his *Methodus Differentialis* (1730), following a proof of Taylor's theorem, Stirling wrote:[6]

Here we have an idea of the analogy between the differential method [*Methodum Differentialem*] and the ordinary method of series: in the latter fluxions, or ultimate ratios of differences, are employed; in the former, in a much more general way, we employ differences of any magnitude. (Stirling (1730), p. 103)

Taylor and Stirling were motivated to adopt this approach to the calculus of fluxions because they realized the usefulness of studying the limiting forms of interpolation formulas. Their researches probably suggested to them that the calculus of limits was a general theory which could link Newton's *methodus differentialis* and Newton's *method of fluxions*. Taylor indeed came close to rejecting the use of infinitesimals:

Some people, because that the Fluxions are proportional to the nascent Increments of Quantities, have thought that by the Method of Fluxions Sir Isaac Newton has introduced into Mathematicks the Consideration of infinitely little Quantities; as if there were any such thing as a real Quantity infinitely little. But in this they are mistaken, for Sir Isaac does only consider the first or last Ratio's of Quantities, when they begin to be, or when they vanish, not after they are become something, or just before they vanish; but in the very moment when they do so. In this case Quantities are not consider'd as infinitely little; but they are really nothing at the time that Sir Isaac takes the Proportions of their Fluxions; and the Truth of this Method is demonstrated from the Principles of the Method of Increments, in the same manner as the Ancients demonstrated their Conclusions in the Method of Exhaustions, by a *Deductio ad Absurdum*. (Taylor (1717), pp. 342–3)

This passage by Taylor is quite an exception: generally early fluxionists did not distinguish between the two approaches to the calculus, one in terms of infinitesimals, the other in terms of limits. The *Commercium Epistolicum* itself ended with the conclusion that

the Differential Method is one and the same with the Method of Fluxions, excepting the Name and Mode of Notation; Mr. Leibniz calling those Quantities Differences, which Mr. Newton calls Moments or Fluxions; and marking them with the Letter d, a Mark not used by Mr. Newton. (Collins (1713), pp. 121–2)

This was the accepted view when Berkeley wrote *The Analyst*.

3.3 The doctrine of prime and ultimate ratios

Berkeley's criticisms provoked much interest in the foundations of the fluxional calculus. In the years immediately following the publication of *The Analyst* (1734) several pamphlets were published in answer to Berkeley, and several periodicals devoted large sections to the debate on foundations.[7] British mathematicians were stimulated to read Newton's work more carefully: different translations of the first eleven lemmas of book I, of the second lemma of book II of the *Principia* (1687), and of the introduction to 'De quadratura' (1704c) were compared.[8] Even though the problems raised by Berkeley remained unsolved, the accuracy in the use of terms greatly improved.

The starting point for the fluxionists' answers to Berkeley was the warning placed after the eleven lemmas of the *Principia* (Book I, Section 1) on the doctrine of prime and ultimate ratios:

Therefore if hereafter I should happen to consider quantities as made up of particles, or should use little curved lines for right ones, I would not be understood to mean indivisibles, but evanescent divisible quantities; not the sums and ratios of determinate parts, but always the limits of sums and ratios; and that the force of such demonstrations always depends on the method laid down in the foregoing Lemmas.[9]

The common answer of the fluxionists was that Berkeley's logical criticism was applicable only to the differential method, which was employed by Newton merely to abbreviate the proofs. Newton's genuine method was the method of limits, or of 'prime and ultimate ratios', which was quite different from the Leibnizian method. In the calculus of limits the whole calculation represents the steps of a process of approximation, and, it was maintained, there is no contradiction in making a quantity equal to zero at a certain point. Therefore, Berkeley was unfair when he attributed the same logical fallacy to the two methods.[10]

However, the fluxionists' theory of limits was not without its difficulties. Newton himself felt the necessity to give some explanations. We read in the *Principia*:

Perhaps it may be objected, that there is no ultimate proportion of evanescent quantities; because the proportion, before the quantities have vanished, is not the ultimate, and when they are vanished is none. But by the same argument it may be alleged that a body arriving at a certain place, and there stopping, has no ultimate velocity; because the velocity, before the body comes to the place, is not its ultimate velocity; when it has arrived, there is none.[11]

Newton went on to give kinematical illustrations of the existence of a limiting value and to explain that his method did not imply the existence of infinitesimals.

These pages appealed to the fluxionists for two reasons. First of all, Berkeley's ontological criticism had received, in their opinion, a full answer: the calculus did not deal with fictional or non-existing entities, but rather with 'tangible' motions and velocities.[12] Secondly, the fallacy of 'shifting a supposition' was implied in the principle of cancellation of higher order differentials but not in the limit processes.

The first point does not appear to be satisfactory. There is a circularity in basing the calculus, a mathematical tool devised to study kinematics, on the concepts of time and velocity. It is interesting to see that this was never really felt to be a difficulty by Newtonian mathematicians. For instance, William Emerson (1701–82) expressed ideas widely shared amongst the British when he wrote:

Let a hollow Cylinder be filled with Water, and let it flow freely out through a Hole at the Bottom of it. It is well known, that the velocity of the effluent Water depends on the Height of the Water within the Cylinder; and therefore, since the Surface of the incumbent Water continually descends without any the least Stop, the Velocity of the effluent Stream will continually decrease, till it all be run out. Therefore it is plain, there can be no two moments of Time, succeeding each other so nearly, wherein the Velocity of the running Water is precisely the same. And therefore the Velocity that the effluent Water has at any given Point of Time, belongs only to that one particular, indivisible Moment of Time, and no other: And this is accurately the fluxion of the Fluid flowing out at that Moment of Time. Now if precisely at that Moment you begin and continue to pour more Water into the Cylinder, so that the surface of the Water may descend no lower, but keep its Place; then the effluent Water will also retain its Velocity, and continue to be the Fluxion of the Fluid as before. Now these are the genuine Effects and Operations of Nature itself, and do, in a manner visibly, confirm the truth of what has been said of the Nature of *Fluxions*. (Emerson (1743), pp. ix–x)

The second point created a lot of problems. As Berkeley noticed, if the ultimate ratio was not a ratio of infinitesimals, it was to be understood as 0/0. Freeing the calculus from infinitesimals opened up the question about the paradoxical nature of the limiting ratio. In fact, Berkeley's opponents gave conflicting explanations of Newton's theory of limits.

An example of alternative interpretations is given by the quarrel between James Jurin (1684–1750) and Benjamin Robins (1707–51).[13] Berkeley was probably amused to see that two fluxionists were unable to agree on such an important subject. Robins and Jurin discussed at great length Newton's theory of limits.[14] They disagreed on the meaning of the phrase 'ultimo fiunt equales' occurring in the first lemma (Book I, section 1) of the *Principia*. According to Robins the limit to which a variable tends is never achieved; his definition was:

we shall in the first place define an ultimate magnitude to be the limit, to which a varying magnitude can approach within any degree of nearness whatever, though it can never be made absolutely equal to it. (Robins (1735), p. 53)

While Jurin wrote:

By arriving at a limit I understand Sir Isaac Newton to mean, that the variable quantity, or ratio, becomes absolutely equal to the determinate quantity, or ratio, to which is supposed to tend. (*The Present State of the Republick of Letters*, XVI (Sept. 1735), p. 300)

The interest of the opposition between Jurin and Robins consists in the fact that they employed different models for the calculus. Robins was able to maintain his limit-avoiding interpretation because he understood limit processes as non-kinematical approximations. He expressly distinguished between a method of fluxions where quantities are considered as varying in time and a method of prime and ultimate ratios, equivalent to the Greek method of exhaustion. In the latter the concepts of time and velocity are not employed. It may be that Jurin adhered more strictly to the *Principia* and conceived the limit processes as kinematical; for instance, in his opinion, the chord and the tangent of Lemma VII (Book I, Section 1) 'at the same instant of time, arrive at the same proportion of a perfect equality' (see *The Present State of the Republick of Letters*, XVI (Sept. 1735), p. 379). Even though the problem of the double meaning of Newton's ultimate ratios was not resolved either by Jurin or by Robins, their discussion had the merit of giving expression to the opposition between two conflicting interpretations of Newton's theory of prime and ultimate ratios.[15]

Another example of logical analysis of the theory of prime and ultimate ratios is given by the *Introduction to the Doctrine of Fluxions* (1736) written by Thomas Bayes (1702–61), one of the fathers of probability theory. As G. C. Smith has shown in (1980), Bayes developed a systematic theory on the operations with limits which resemble those of Cauchy. He stated the following laws:

(1) If ult.$a:b = A:B$ and ult.$b:d = B:D$ then ult. $a:d = A:D$, where ult. means 'the ultimate ratio of'.

(2) If in a time interval T for the fluents a, b, x and y holds the proportion: $a:b::A \pm x:B \pm y$, and at the end of that time a, b, x and y vanish then: ult.$(a \pm b)/b = (A \pm B/B)$.

By using (1) and (2) Bayes showed that $\dot{x}/\dot{y} = $ ult.$(\Delta x/\Delta y)$, that the fluxion of $a+b$ is equal to $\dot{a}+\dot{b}$ and that the fluxion of the product ab is equal to $a\dot{b}+\dot{a}b$. However, Bayes was obviously very far from understanding the concept of function and his theorems are always referred to geometric processes of approximation. Henry Pemberton, Jacob Walton and James Smith also had a minor role in the controversy with Berkeley. But one can safely say that the controversy did not end in 1735: in all the treatises on fluxions of the second half of the century there was a preface dealing with the vindication of Newton's theory against the criticisms of the Bishop of Cloyne.

3.4 The foundations of the calculus in Maclaurin's *Treatise of Fluxions*

The most authoritative answer to Berkeley was given by Colin Maclaurin. In the *Treatise of Fluxions* (1742) he presented an interpretation of Newton's calculus which had a great influence in Great Britain.[16] With few exceptions, Maclaurin's view of the calculus dominated the second half of the century: the infinitesimals and the moments of the early fluxionists were definitively abandoned. Maclaurin (1742) begins with the following declaration:

A LETTER published in the year 1734, under the title of *The Analyst*, first gave occasion to the ensuing Treatise, and several reasons concurred to induce me to write on this Subject at so great a length. The Author of that Piece had represented the method of Fluxions as founded on false Reasonings, and full of Mysteries. His Objections seemed to have been occasioned, in a great measure, by the concise manner in which the Elements of this Method have been usually described; and their having been so much misunderstood by a person of his abilities, appeared to me a sufficient proof that a fuller Account of the Grounds of them was requisite.

Though there can be no comparison made betwixt the extent and usefulness of the antient and modern Discoveries in Geometry, yet it seems to be generally allowed that the Antients took greater care, and were more successful in preserving the Character of its Evidence entire. This determined me, immediately after that Piece came to my hands, and before I knew any thing of what was intended by others in answer to it, to attempt to deduce those Elements after the manner of the Antients, from a few unexceptionable principles, by Demonstrations of the strictest form. (Maclaurin (1742), pp. vii–viii)

The Preface is followed by a long introduction on the method of exhaustion. Maclaurin's purpose is to show that the calculus of fluxions is a generalization of the 'geometry of the antients'. He clearly opposed the 'method of the antients' and the method of fluxions to the method of infinitesimals:

In what Archimedes had demonstrated of the limits of figures and progressions, there were valuable hints towards a general method of considering curvilinear figures; so as to subject them to mensuration by an exact quadrature, an approximation, or by comparing them with others of a more simple kind. Such methods have been proposed of late in various forms, and upon different principles. The first essays were deduced from a careful attention to his steps. But, that his method might be more easily extended, its old foundation was abandoned, and suppositions were proposed which he had avoided. It was thought unnecessary to conceive the figures circumscribed or inscribed in the curvilinear area, or solid, as being always assignable and finite; and the precautions of Archimedes came to be considered as a check upon Geometricians, that served only to retard their progress. Therefore, instead of his assignable finite figures, indivisible or infinitely small elements were substituted; and these being imagined indefinite, or infinite, in number, their sum was supposed to coincide with the curvilinear area, or solid. (Maclaurin (1742), p. 37)

Maclaurin's opposition to the method of infinitesimals, to the 'Geometricians' who 'had involved themselves in the mazes of infinity' occupies many pages: they, who abandoned the 'constant practice of the antients', have accepted the infinite divisibility of matter and the vortex theory of planetary motions:

From geometry the infinites and infinitesimals passed into philosophy, carrying with them the obscurity and perplexity that cannot fail to accompany them. An actual division, as well as a divisibility of matter *in infinitum* is admitted by some. Fluids are imagined consisting of infinitely small particles, which are composed themselves of others infinitely less; and this sub-division is supposed to be continued without end. Vortices are proposed, for solving the phaenomena of nature, of indefinite or infinite degrees, in imitation of the infinitesimals in geometry; that, when any higher order is found insufficient for this purpose, or attended with an insuperable difficulty, a lower order may preserve so favourite a scheme. Nature is confined in her operation to act by infinitely small steps. Bodies of a perfect hardness are rejected, and the old doctrine of atoms treated as imaginary, because in their actions and collisions they might pass at once from motion to rest, or from rest to motion, in violation to this law. Thus the doctrine of infinites is interwoven with our speculations in geometry and nature. (Maclaurin (1742), p. 39)

However, we have already found Maclaurin using infinitely little quantities in his *Geometria Organica* (1720c) (see chapter 2, section 2.3). His shift from a differentialist approach to the rejection of infinitesimals is quite representative of the change which occurred after Berkeley's *Analyst* (1734). In a letter to Stirling dealing with the preparation of the *Treatise of Fluxions* (1742), Maclaurin confessed:

I am not at present inclined to put my name to it [the *Treastise of Fluxions*]. Amongst other reasons there is one that in my writings in my younger years I have not perhaps come up to that accuracy which I may seem to require here. When I was very young I was an admirer too of infinites; and it was Fontenelle's piece that gave me a disgust of them or at least confirmed it together with reading some of the Antients more carefully than I had done in my younger years. (in Tweedie (1922), p. 74)

How then is it possible to link 'Archimedes's method' with the method of fluxions, avoiding the unsafe concept of infinitesimal? Maclaurin states his programme very clearly:

In explaining the Notion of a Fluxion I have followed Sir Isaac Newton in the first Book, imagining that there can be no difficulty in conceiving Velocity wherever there is Motion; nor do I think that I have departed from his Sense in the second

Book; and in both I have endeavoured to avoid several expressions, which, though convenient, might be liable to exceptions, and, perhaps, occasion disputes. I have always represented Fluxions of all Orders by finite Quantities, the Supposition of an infinitely little Magnitude being too bold a *Postulatum* for such a Science as Geometry. But because the Method of Infinitesimals is much in use, and is valued for its conciseness, I thought it was requisite to account explicitly for the truth, and perfect accuracy of the conclusions that are derived from it. (Maclaurin (1742), p. x)

In the first chapter Maclaurin presents the kinematical model of fluxions. Motion, space and velocity are given as non-problematic primitive notions. The fluxion is defined as 'the velocity with which a quantity flows, at any term of the time while it is supposed to be generated' (Maclaurin (1742), p. 57). The reference to the kinematical model cannot be analysed any further: it is this intuitive reference that, according to Maclaurin, gives an ontological basis to the calculus. However, in order to reduce the calculus to Archimedean geometry, it is necessary to give a geometrical representation of the kinematical model of fluxions. Maclaurin tries to associate with the velocity a finite proportional geometrical quantity employing the following operational definition:

The velocity with which a quantity flows, at any term of the time while it is supposed to be generated, is called its *Fluxion*, which is therefore always measured by the increment or decrement that would be generated in a given time by this motion, if it was continued uniformly from that term without any acceleration or retardation. (Maclaurin (1742), p. 57)

This is equivalent to defining the 'measure' of $y'(t)$ as $y'(t)\Delta t$: a procedure that involves circularity.[17] Nonetheless, this definition was quite common in the seventeenth and early eighteenth centuries.

After these definitions Maclaurin introduced four axioms (two on accelerated and two on 'retarded' motion) which expressed the fundamental properties of motion and velocity:

Axiom I

The space described by an accelerated motion is greater than the space which would have been described in the same time, if the motion had not been accelerated, but had continued uniform from the beginning of the time.

Axiom II

The space described by a motion while it is accelerated, is less than the space which is described in an equal time by the motion that is acquired by that acceleration continued uniformly.

Axiom III

The space described by a retarded motion is less than the space which would have been described in the same time, if the motion had not been retarded, but had continued uniform from the beginning of the time.

Axiom IV

The space described by a motion while it is retarded, is greater than the space which is described in an equal time by the motion that remains after that retardation, continued uniformly. (Maclaurin (1742), p. 59)

While the operational definitions provide a geometrical representation of kinematical magnitudes, the axioms allow the inequalities between the kinematical magnitudes to be formulated.[18] All the proofs in the first book of Maclaurin's *Treatise of Fluxions* (1742) resemble in structure the method of exhaustion in so far as they prove the impossibility of an inequality between two geometrical magnitudes. Maclaurin is extremely prolix and sometimes almost unreadable because he compels himself to adhere strictly to the indirect geometric proofs even when he is treating advanced topics. Only in the second book is the 'geometry of fluxions' abandoned in favour of the 'calculus of fluxions'. Here Maclaurin introduces the notation and the algorithm of Newton's 'De quadratura' (1704c). But now we know that it would be possible to retranslate all the theorems into the kinematic style of the first book:

The evidence of the method had been disputed, and objections had been made to the number of symbols employed in it, as it might serve to cover defects in the principles and demonstrations. In order to obviate any suspicions of this kind, we endeavoured to describe it in a manner that might represent the theorems plainly and fully, without any particular signs or characters, that they might be subjected more easily to a fair examination. (Maclaurin (1742), p. 575)

Maclaurin always refers the reader to articles in the first book, in order to show that the symbolic results of the calculus are interpretable in terms of kinematical magnitudes. For instance, when in the second book he proves Taylor's theorem (pp. 600–11) he refers the reader to article 255 where he tried to prove by geometry that in the increment of a fluent can be 'distinguished' a 'part which measures the first fluxion', '$\frac{1}{2}$ of that which measures the second fluxion of the ordinate, $\frac{1}{6}$ of that which measures its third fluxion, $\frac{1}{24}$ of that which measures its fourth fluxion, and so on'.

Maclaurin's foundation of the calculus of fluxions on the kinematical model of fluxions met with a very favourable reception amongst British mathematicians. Criticisms of Maclaurin concerned the tediousness of his style and nobody tried to reproduce the lengthy proofs of the *Treatise of Fluxions* (1742): however, since 1742 almost all the fluxionists accepted Maclaurin's rejection of infinitesimals.

Maclaurin's *Treatise of Fluxions* systematized ideas deeply rooted in British science. First of all, it made it possible to understand the unity which existed between the branches of mathematics: there was no need to choose between conflicting methods. Everything could be reduced to a common field: a kinematic geometry based upon our intuition of motion, time and velocity. Secondly, Maclaurin's *Treatise of Fluxions* was in accord with the classicism of British scientists; it was often repeated that the 'new analysis' was just a generalization of 'Archimedes' method', but nobody had tried to show in detail how it was possible to apply the 'antient geometry' not only to the determination of the tangent of a parabola but to the brachistocrone or to Taylor's theorem. Lastly, the need to provide an objective reference to the calculus was satisfied. The theorems of the calculus did not deal with 'fictions' or 'ghosts of departed quantities' but had a kinematic meaning: as Newton wrote in 'De quadratura' (1704c) the fluents and fluxions 'have an existence in nature'.[19]

PART II

THE MIDDLE PERIOD

4

THE TEXTBOOKS ON FLUXIONS
(1736–58)

IN THE two decades which followed the dispute with Berkeley the calculus of fluxions was given shape in a number of 'treatises'. We have already looked at the best achievement of this generation of textbooks, Maclaurin (1742). In this chapter I have grouped together the other treatises. The first thing which strikes one about them is their quantity rather than their quality. It has been possible to estimate that from 1736 to 1777 about 18 000 copies of treatises on fluxions were sold in Great Britain. This situation contrasts sharply with the first three decades of the century, in which the calculus of fluxions was known to very few mathematicians. It was about the middle of the century that the world of 'philomaths' and, perhaps, some students in the military academies and universities began to practise with Newton's dots. The form the calculus would take in the second half of the eighteenth century very much depended upon the period covered in this chapter.

4.1 Teaching the algorithm of fluxions

The flood of controversial pamphlets on foundations published in the years 1734–6 (see chapter 3) was followed by an intense production of 'textbooks' on the calculus of fluxions. They were intended for beginners, as Hayes (1704), Ditton (1706) and Stone (1730) had been: the reader was introduced to the notation, rules and application of the calculus. The only requisite was a knowledge of algebra (but not of series) and geometry (with or without trigonometry). It is remarkable that almost all the textbooks on fluxions were published between 1736 and 1758. These works completely superseded their early eighteenth-century predecessors (with the exception of course of Newton's 'De quadratura' (1704c) and 'De analysi' (1711b)). The teaching and the 'image' of the calculus were to a great extent shaped in these twenty-two years. It is through the treatises on fluxions published in this period that Newton's mathematical work was reinterpreted and systematized: as a result it is possible to identify in the second half of the eighteenth century a fluxional school which based its teaching on an homogeneous set of textbooks.

In fact from 1736 to 1758 twelve textbooks on fluxions were published: those by Hodgson (1736), Muller (1736), Newton (1736), Simpson (1737) and (1750c), Blake (1741), Maclaurin (1742), Emerson (1743), Rowe (1751), Rowning (1756), Saunderson (1756a), Lyons (1758). In addition, Martin (1739) and Simpson (1752) had chapters on fluxions.[1] Some of these were reissued several times during the eighteenth century and up to the 1820s. After 1758 only three new treatises on fluxions were published: Holliday (1777), an elementary introduction which was not reissued; Vince (1795), which was an editorial success; and the sloppy Dealtry (1810). Hutton (1798, 1801) had a section on fluxions, which was extended in a third volume for the sixth edition of 1811 with the assistance of Olynthus Gregory.

The Doctrine of Fluxions (1736) by James Hodgson (1672–1755) was one of the first works intended for teaching in which the programme, derived from the Berkeley controversy, of distinguishing between the differential and the fluxional calculus was adopted. Hodgson wrote:

The Design of publishing the following Treatise, is to introduce the true Method of *Fluxions*, most of the Books that have hitherto appeared upon that Subject having in them little more than the Name, the Principles upon which they have proceeded being the same with the *Differential Calculus*; so that by calling a *Differential* a *Fluxion*, and a second *Differential* a second *Fluxion*, &c. they have so confusedly jumbled the Methods together, that People, who have not been thoroughly acquainted with them, have been led into many Mistakes: For although the way of Investigation in each be the same, and both center in the same Conclusions, yet whoever will compare the Principles, upon which the Methods are founded, will find that they are very different. (Hodgson (1736), 2nd edn., p. v)

Hodgson kept to Newton's 'De quadratura' (1704c) as, in his view, it was the work where the 'true Method of Fluxions' was to be found: it is ironic that in the same year a treatise by Newton appeared in which infinitesimals were employed.

This treatise was *The Method of Fluxions and Infinite Series* (Newton (1736)), a translation by John Colson (1680–1760) of a manuscript in Latin written in about 1671. It was often lamented that the world had had to wait so many years to see Newton's masterpiece on fluxions. It is astonishing to realize that publication sixty years beforehand would have changed the history of the calculus and would have avoided for Newton any controversy over priority. In 1736 all the results contained in

Newton's treatise were well known to mathematicians. However, it was too concise for a beginner, and Colson added almost 200 pages of explanatory notes. His commentary contributed to the establishment of a kinematical approach to the problem of foundations. In his explanatory notes Colson presents the 'geometrical and Mechanical Elements of Fluxions'. He writes:

The foregoing Principles of the Doctrine of Fluxions being chiefly abstracted and Analytical, I shall here endeavour, after a general manner, to shew something analogous to them in Geometry and Mechanicks; by which they may become not only the object of the Understanding, and of the Imagination, (which will only prove their possible existence) but even of Sense too, by making them actually to exist in a visible and sensible form. (Newton (1736), p. 266)

Colson was convinced that by using moving diagrams it is possible to exhibit 'Fluxions and Fluents Geometrically and Mechanically [...] so as to make them the objects of Sense and ocular Demonstration' (Newton (1736), p. 270). The motivation for using the geometrical and mechanical elements of fluxions is clearly that of guaranteeing an ontological basis to the calculus; in fact:

Fluents, Fluxions, and their rectilinear Measures, will be sensibly and mechanically exhibited, and therefore must be allowed to have a place *in rerum natura*. (Newton (1736), p. 271)

Colson's approach to the calculus is representative of a whole generation of British mathematicians: his 'sensibly exhibited rectilinear measures' of fluxions are a naive anticipation of Maclaurin's kinematic definitions of the basic concepts of the calculus.

The need to render Newton's original works accessible to 'beginners' was also the motive behind John Stewart's (d.1766) translation and commentary of 'De analysi' (1711b) and 'De quadratura' (1704c) published in Newton (1745). It seems that the English editions of Newton's works on fluxions turned out to be profitable, since in 1737 there appeared Newton (1737), a pirated edition of the manuscript 'de methodis', already translated by Colson as Newton (1736).[2] A *Mathematical Treatise* by John Muller (1699–1784) was published in 1736 and translated into French in 1760. A first part devoted to conic sections was followed by a fairly simple textbook on the application of the calculus to the study of 'curves' (including maxima, minima, curvature) and to mechanics (centres of gravity and oscillations, pendulums, projectiles) but not to geodesy or physical astronomy. Another attempt to popularize fluxions was *An Explanation of Fluxions* (1741) by Francis Blake (1708–80). This short work was highly regarded for its rigour. Simpson declared his debt to Blake in the Preface to his treatise (1750c) and the *Explanation* was reprinted in 1763 and 1809.

The treatises reprinted most were William Emerson's *The Doctrine of Fluxions* (1743; 2nd edn. 1757; 3rd edn. 1768; 4th edn. 1773), Thomas Simpson's *The Doctrine and Application of Fluxions* (1750c; 2nd edn. 1776; 3rd edn. 1805; 4th edn. 1823) and John Rowe's *An Introduction to the Doctrine of Fluxions* (1751; 2nd edn. 1757; another? 1762; 3rd edn. 1767; 4th edn. 1809). Up to the beginning of the nineteenth century these were studied more than Maclaurin's *Treatise of Fluxions* (1742; 2nd edn. 1801), which cannot be considered a textbook for beginners, and perhaps even more than the works of Newton.

Simpson (1750c) is certainly the most advanced. It is divided into two volumes. The first one covers the subjects usually found in the other treatises and can be considered by itself as an elementary treatise (the structure should be predictable: a chapter on foundations, one on notation and the rules of differentiation and integration by power series, and a sequel of applications to geometry and mechanics). The second volume goes much further: of particular value are section v on Cotes's integrals and section x on the attraction of spheroids (see appendix A.2). Emerson (1743) also contains advanced subjects (see appendix A.1): in 300 pages it covers a great variety of problems.[3] Rowe (1751) tried to supply an easier work. He confined himself to the simplest applications of the calculus (see appendix A.3). In the third and last part, devoted to the solution of 'miscellaneous questions', Rowe refers the reader for more advanced problems to Hayes (1704), Ditton (1706), Stone (1730), Hodgson (1736), Muller (1736), Newton (1736, 1737), Maclaurin (1742), Emerson (1743), Newton (1745), Simpson (1750c) (and in the third edition published in 1758 also to Saunderson (1756a) and Lyons (1758)).

The treatises by Rowning (1756), Saunderson (1756a) and Lyons (1758) concluded the period taken into consideration in this chapter. We have already come across Saunderson's treatise, which was read in Cambridge in the 1730s. Rowning's and Lyons's treatises answered the same purpose as Rowe (1751), that of simplifying Emerson (1743) and Simpson (1750c).

All these treatises, however advanced they may have been, did not introduce the student to the calculus as a theory (this was generally the case in, for instance, treatises on plane geometry or conic sections), but rather explained to him how to employ in geometry and mechanics a set of rules established in a first concise section. The only exception is Colin Maclaurin's *Treatise of Fluxions* (1742) which, as we know, was so deep on the 'kinematical axiomatization' of the calculus and on methodological questions. This peculiarity of Maclaurin's work was generally criticized as a drawback for teaching purposes and as the cause of its failure on the market for science books. For instance, we find Francis Blake writing to Thomas Simpson in 1741 about his forthcoming *An Explanation of Fluxions* (1741):

But should not we wait, think you, for a Sight of Mr. Maclaurin's Work, which you say he is about publishing? I have no Apprehension that he will interfere with me for the Reason given in my Introduction; beside I remember to have seen several Sheets of it, long ago, and if he has not altered his method, I dare say he addresses himself only to great Mathematicians. He begins with a tedious Account of Archimedes's Manner of Reasoning in Mathematics; after which he lays down a vast Nr. of Lemmas concerning the Nature of Motion; and the Method of Demonstration used throughout is (as well as I can judge) an exact copy of the ancients.[4]

The reply from Simpson was encouraging:

I am confident that a much plainer book may be wrote on ye Subject then his [Maclaurin's], and that an easy Explication of ye Theory [...] will be of much greater use to a beginner than both those large volumes.[5]

In fact Blake and Simpson were right in predicting that Maclaurin's technical *Treatise of Fluxions* (1742) could not meet the demand of easier presentations of Newton's calculus. In 1743 Maclaurin, writing to Sir Andrew Mitchell, observed:

I see to day a treatise of fluxions advertised 5 sh. in sheets, the author not named, sold by Wm Innys [probably Emerson (1743)]. I fear my book will hardly clear itself. (Colin Maclaurin to Sir Andrew Mitchell, 31 March 1743, in Maclaurin (1982), p. 101)

After a month he added:

I am not surprised that my book has not sold fast, and my Ambition or expectation as to that is not high. I have a sum to pay the printers here still of about 80 L. which they are pushing for, but I am to postpone the payment, if I can at a meeting we are to have to morrow. (Colin Maclaurin to Sir Andrew Mitchell, 5 April 1743, in Maclaurin (1982), p. 102)

Even though Maclaurin's geometric method was not always adopted, his foundation of the calculus on kinematics was accepted by the other textbook writers. In all these textbooks the reader was introduced in a preface or first chapter to the kinematic meaning of the concepts of the calculus: here Maclaurin was followed as the authority on these foundational aspects. Then the calculus of fluxions was briefly treated as a set of simple rules (basically the rule of differentiation of xy, the chain rule, the integration of x^n and of $x+y$ and integration by parts). The notation was strictly that of Newton. Even though Newton's approach might have suggested the concept of a functional dependence of variable quantities upon time, a functional notation (such as $f(t)$) was not explicated. The rest of the textbook was taken up with the applications of the calculus.

Fluxional textbooks were analytical in the sense that they were all concerned with the use of an algorithm, which was however safely grounded on kinematical concepts. The calculus of fluxions was never presented in the treatises as a purely mathematical subject: it was not conceived as a universal language (like algebra) or as a systematic theory (like geometry). The calculus was the language of continuously varying magnitudes and was immediately applied to the study of the universe of flowing quantities. In a textbook on fluxions there were only two purely mathematical chapters: one on series and another on fluxional equations. Series were conceived as a part of algebra and they were included in a fluxional treatise because all the techniques of integration (with the exception of Cotes's integrals which were sometimes used) depended upon power series. Even in the best textbooks, such as Simpson (1750c), fluxional equations did not receive a decent treatment. Rather than being given a general theory of integration, like the one to be found in Euler's *Istitutiones Calculi Integralis* (1768–70), the reader of Simpson (1750c) had just fifteen disconnected examples of first order fluxional equations to study.

However, the level of the textbooks on fluxions was not representative of the contemporary researches on foundations of Bayes, Robins and Jurin (see chapter 3, section 3.3) and of the researches of Maclaurin and Simpson (see chapter 5). The reasons why the level of textbooks was so much poorer than that of research works are to be found in a study of the relationships between textbook writers and their readers and editors. It seems that in the 1730s and 1740s there was an increase in the number of potential readers of introductory works on the Newtonian calculus who were, however, not in the position to appreciate such an advanced treatment as that offered by Newton or Maclaurin.

4.2 Textbooks on fluxions and the science books trade

Several of the textbooks mentioned in the last section were reissued. In particular between 1736 and 1777 we have calculated that twenty-four editions of textbooks were completely devoted to the calculus of fluxions.[6] The problem naturally arises of understanding the reasons for this increasing interest in a quite technical aspect of Newtonian science. In order to find a plausible explanation I will concentrate on the authors of treatises on fluxions, their readers, and their editors.

The main characteristic which distinguished the authors of the treatises on fluxions published in the period 1736–58 from their predecessors, such as Hayes or Stone, was that they were connected in one way or another with universities or military academies: that is, they were professors of mathematics or they worked merely as tutors or coaches. The only exceptions were Francis Blake (1708–80) and William Emerson (1701–82). Blake was an *amateur*, perhaps a student of Thomas Simpson, who devoted himself to natural philosophy: he contributed two papers (1753) and (1760) on the steam-engine in the *Philosophical Transactions* and was elected FRS in 1746. Emerson was taught at Newcastle and York, but, we are told in Bowe (1793), he learnt mathematics by himself when he was over thirty years old. He dedicated himself to teaching for a short period and with little success in Hurworth. Emerson was a professional textbook writer: he had a contract with John Nourse, the famous publisher, for whom he wrote several works on mechanics, mathematics and astronomy.[7]

As we already know, Maclaurin was Professor of Mathematics at Edinburgh University, while Thomas Simpson taught at Woolwich.[8] Thomas Simpson (1710–61), the son of a weaver in Leicester, was one of the many self-taught mathematicians in the eighteenth century. At the beginning of his career he settled in Nuneaton, a village close to his birthplace, where he earned money as a fortune teller. His career was interrupted by an accident, and he was compelled to hide for a while. We find him, about 1735, in London married to an older woman. Simpson gained a reputation as a mathematician by answering some questions in the *Ladies' Diary* and by publishing *A New Treatise of Fluxions* (1737). In 1743 he was appointed Master of Mathematics at the newly formed Royal Military Academy at Woolwich. The list of his works is impressive. The most important are *The Doctrine and Application of Fluxions* (1750c), *Mathematical Dissertations* (1743) and *Miscellaneous Tracts* (1757). The first is one of the best textbooks on fluxions, even though not comparable with Maclaurin (1742). The second contains a remarkable treatment of the attraction of spheroids, while the third is mainly concerned with physical astronomy: the precession of equinoxes, the orbit of comets, the motion of the Moon's apogee.

Muller (1699–1784) too was at Woolwich, from 1741 as Chief Master (later as Professor of Fortification and Artillery) of the Royal Military Academy. Francis Holliday (1717–87) was head-master at Houghton Park, Retford. Nicholas Saunderson (1682–1739) and John Colson

(1680–1760) occupied the Lucasian Chair at Cambridge. John Colson, before his appointment in Cambridge, was teacher of mathematics of Sir Joseph Williamson's Mathematical School at Rochester. His concern with mathematical education is testified by his commentary of Newton (1736). He also contributed to the edition of Saunderson (1740) and began a translation (1801) of Agnesi's *Istituzioni Analitiche*.[9] John Rowning (1700–71) was educated in Cambridge; he gave lectures on natural philosophy in Cambridge and London, and was head-master of the Spalding Gentleman's Society from 1759 to 1771. In the 1730s he began publishing the first parts of a successful *Compendious System of Natural Philosophy* (1744, 1745). Israel Lyons (1739–75) lived in Cambridge, Oxford and London. In 1762–4 he delivered in Oxford a course of lectures on botany. His knowledge of mathematics was appreciated, since he worked several times for the Nautical Almanac and, as a cartographer, for the Board of Longitude. In 1773 he was the astronomer in the polar expedition of C. J. Phipps. John Stewart (d.1766) was Professor of Mathematics in Marischal College, Aberdeen, while James Hodgson (1672–1755) was Flamsteed's assistant at Greenwich from 1696 to 1702. He taught mathematics in London, succeeding to John Harris (see chapter 1, section 1.2) at the Marine Coffee House, and in 1709 he became Master of the Royal Mathematical School at Christ's Hospital.[10]

For whom were the treatises on fluxions intended? The first and most natural answer is that Simpson's or Colson's readers were their students, or students in other academies or universities. A confirmation of this is the title of Simpson's *Select Exercises for Young Proficients in the Mathematicks* (1752), a widely used slim volume of exercises in which we also find a short treatment of fluxions; more ambiguously, in the preface Simpson gives methodological instructions to 'the learners'.[11] The mathematical treatises which appeared around the middle of the century were generally addressed to 'learners' or 'beginners'. Only with Vince (1795) and Dealtry (1810) do we have treatises on fluxions 'designed for the use of students in the Universities'.

Muller indicated in the title pages of his *Treatise Containing the Elementary Part of Fortification* (1746), *Elements of Mathematics* ·(1748), *Treatise Containing the Practical Part of Fortification* (1755) and *Treatise of Artillery* (1757) that these were 'for the Use of the Royal Academy of Artillery'. But students in the Royal Naval Academy at Portsmouth, in Woolwich, and later in Sandhurst numbered only a few hundred (see chapter 8). Muller's and Simpson's successful publishing careers could not have been based on their students. Furthermore, it does not seem plausible that fluxions were taught in an eighteenth-century British military academy. The most that could be achieved, as is well shown by the output of Charles Hutton (1737–1823), one of the professors at Woolwich, was practical engineering and ballistics.[12] Indeed, Muller predicted in the preface to his *Treatise Containing the Practical Part of Fortification* (1755), see p. vii, that some of his readers 'may not understand algebra'. It is therefore probable that the reference to the Royal Military Academy in Muller's works had the role of giving his volumes a prestigious gloss.

A more interesting market was to be found in Cambridge. With the establishment in the 1740s of the Senate House Examination, the system of selecting students at Cambridge was given a clear mathematical basis. It is difficult to trace the history of what was to become the Tripos Exam. The first Tripos lists were published in 1747. It seems that, at the beginning, the exam was oral, and it was only in the 1770s that written answers were required. However, it was almost completely a mathematical exam, and we have evidence that students were asked to solve problems of mechanics and fluxions.[13]

This situation was not representative of other universities (especially Oxford where mathematics was hardly touched on); it became a peculiarity of Cambridge which was regretted by some:

Sir,——— College, Cambridge, June 30, 1756

Mathematics is the standard, to which all merit is referred; and all other excellencies, without these, are quite overlooked and neglected: the solid learning of Greece and Rome is a trifling acquisition; and much more so, every polite accomplishment: in short, if you will not get all Euclid and his diagrams by heart, and pore over Saunderson 'till you are as blind as he was himself, they will say of you, as in the motto of one of your late papers, actum est! ilicet! peristi! 'tis all over with you! you are ruined! undone! [...] And, indeed, there seems to be a strong analogy between the inclemency of the weather attacking our bodies, and the storm of afflictions which batter our minds.[14]

So 'B.A.' wrote in *The Connoisseurs*.

The number of students at Cambridge at that time was under 2000: the maximum figure, approximately 2150, was reached at the beginning and at the end of the century. Some of these students might well have been interested in becoming wranglers: John Jebb, second wrangler in 1757, remembers that for the Tripos the students were requested to solve problems taken from 'the fluxional treatises of Lyons, Saunderson, Simpson, Emerson, Maclaurin and Newton' (Wordsworth (1877), p. 50) In fact these works (with the exception of Maclaurin (1742) and Newton (1736)) were conceived more as exercise texts rather than as treatises: they enabled the reader to apply a concise set of rules to a variety of problems in geometry and mechanics. The fact that the Tripos Exam was introduced in the 1740s and the increase in the publication of mathematical treatises in this same period may be connected. These treatises on fluxions were well suited for students preparing an exam.

It is difficult to say how many students read treatises on fluxions and how advanced the average knowledge of mathematics was. The extant accounts are extremely contradictory: some, like 'B.A.' and Jebb, describe Cambridge as a household of mathematicians, but we have recollections of would-be wranglers who considered the extraction of a square root to three decimal places to be a 'rather severe' test (Wordsworth (1877), p. 52). It is probable that the situation varied: many did not concern themselves with mathematics at all (these many idlers were called 'oi polloi'), but some did. In the last quarter of the century a numerous group of mathematicians was connected in one way or another with Cambridge: e.g. Samuel Vince (1749–1821), James Wood (1760–1839), Edward Waring (1734–98), John Rowning (1700–71), George Atwood (1746–1807), Francis Maseres (1731–1824), Nevil Maskelyne (1732–1811), William Frend (1757–1841), John Brinkley (1763–1835), Henry Cavendish (1731–1810) and Robert Woodhouse (1773–1827). Self taught mathematicians, so typical in the first half of the century, were decreasing, while the group of academically trained mathematicians was growing.

The Scottish universities were another possible market for fluxional treatises. Within the framework of the humanistic education provided in Scotland mathematics may not have been given the same importance as in Cambridge. The adoption of a more harmonious approach meant that mathematics and natural philosophy were taught together with the moral sciences and philosophy. As Richard Olson has shown in his (1971), a bias toward geometrical methods certainly characterized the universities of the Scottish Enlightenment and may have hindered the study of the calculus.[15]

A different group of people who might have been readers of treatises on the calculus is to be found in P. J. Wallis (1976), where 4500 'mathematicians' active between 1701 and 1760 are classified, including mathematics teachers, instrument makers, land surveyors, etc. It would be interesting to know to what extent a mathematical preparation was required for the professions connected with navigation, military and civil engineering, land surveying, cartography, and so on. Rather than a professionalization of mathematics (in Great Britain this process did not take place until the nineteenth century), there may have been a 'mathematization' of several professions.

In eighteenth-century Great Britain there was a widespread interest in mathematics which encompassed also amateurs: the famous 'philomaths'.[16] We have already seen that itinerant mathematics teachers and demonstrators of experiments attracted a numerous public in London and in the provinces at the beginning of the century. This phenomenon was accompanied by the writing of simple introductory works on natural philosophy. As I have already remarked, the calculus of fluxions was too technical a subject to attract the interest of this kind of public. But we have to consider that periodicals such as the *Ladies' Diary* contributed towards extending the interest in mathematics, and as the century progressed the readership of mathematics texts became more numerous. In the 1720s the 'philomaths' gathered together at the Spitalfields Mathematical Society (1717), the Spalding Gentleman's Society (1717), the Manchester Mathematical Society (1718) and the Northampton Mathematical Society (1721). It is known that other mathematical societies existed at York, Lewes, Wapping and Oldham (see Howson (1982), p. 258n).[17] It is interesting to note that fluxions were almost absent from the answers published for the *Ladies' Diary* before 1730. It seems that only in the 1730s the 'philomaths' learnt to use the calculus. The fact that there were these large numbers of 'philomaths', whose delight was puzzle-solving, may well explain why the treatises on fluxions were given the form of exercise texts. This is perhaps the best clue to understanding the increase in output of introductory works on fluxions in the 1730s and 1740s.

It was around the middle of the century that the market for science books reached its highest point of expansion. A general increase in readers of science books, together with a specialization on the part of the *habitués* of coffee houses and mathematical societies, may have convinced some publishers to issue volumes on mechanics, astronomy, instrument making and mathematics.

Of course the great majority of science books were still very simple and unmathematical. Works such as Roger Cotes's *Hydrostatical and Pneumatical Lectures* (1738), Rutherforth's *System of Natural Philosophy* (1748), John Rowning's *Compendious System of Natural Philosophy* (1744, 1745) or Helsham's *Course of Lectures in Natural Philosophy* (1739), which did not require any mathematical preparation, were still the best-sellers. The list of subscribers to Rutherforth (1748), a ponderous work which consisted of lectures in mechanics, optics, hydrostatics and astronomy given in St John's College, Cambridge, totals more than 1000. Cotes and Rowning also lectured in Cambridge, while Helsham taught at the University of Dublin. Whereas at the beginning of the eighteenth century Keill and Whiston could combine theology, mathematics and astronomy, as the century progressed popular lectures on natural philosophy and technical treatises became two completely separate types of science books. The existence of two different levels is revealed by the custom in popular science books of giving the references to technical treatises on mechanics, fluxions or physical astronomy in the footnotes.[18]

The most famous scientific publishers were all interested in publishing both popular and technical science books. The most important was perhaps John Nourse who had a contract with Emerson but also published quite a lot of Simpson's works (see appendix B). These books were expensive: e.g. in 1767 William Emerson's *Method of Increments* (1763) cost about seven shillings, Thomas Simpson's *Doctrine and Application of Fluxions* (1750c) cost twelve shillings, and Newton's *The Method of Fluxions and Infinite Series* (1736) (with Colson's commentary) was fifteen shillings. However, a customer of John Nourse or Millar was probably prepared, for professional reasons or an addiction to science, to spend much more on clocks, telescopes, or air-pumps. We do not have any reliable data on sales figures: however, Feather (1981) found the agreements, correspondence and accounts of John Nourse. From these documents he was able to conjecture that the normal run of a Nourse edition was 1000. We know from Wallis and Wallis's (1986), pp. 221, 317, that 750 copies of Simpson's *New Treatise of Fluxions* (1737) were printed for £1.00; that probably only 350 copies of Stewart's translation of Newton's 'De analysis' and 'De quadratura' (1745) were published, costing £111, and that nearly a half remained in 1747. Mathematical works were, then as nowadays, expensive and we should expect that a run of 1000 would have been excessive. If we take into consideration all the separate printings in the forty years following 1736, considering average runs of 750, we can estimate that in this period about 18000 copies of textbooks on fluxions were sold in Britain.[19]

In conclusion, it seems that the increase in the publication of treatises on the calculus after about 1735 is an aspect of a more general improvement in the market for science books. Both popularizations of natural philosophy and technical works were printed: publishers found it was in their interest to specialize at both ends of the market. The demand for a science education was strong in eighteenth-century Britain. It is difficult to assess the status of those who were interested in mathematical subjects. A small number were located in the universities (Cambridge, Oxford, Glasgow, Aberdeen, St Andrews, Edinburgh, Dublin); the authors of treatises on the calculus were often professors of mathematics, and some of their students might have been stimulated to study the fluxional calculus. In this period a class of academic mathematicians and engineers was being educated in British universities; they were potential consumers as well as producers of science books. But there were also two other large groups who were interested in fluxions: professional men such as cartographers, instrument makers, land surveyors, etc., who needed a knowledge of mathematics, and the philomaths of the *Ladies' Diary*.

5

SOME APPLICATIONS OF THE CALCULUS
(1740–3)

THE PERIOD which followed the death of Newton did not witness only the disputes about foundations: with Maclaurin's *Treatise of Fluxions* (1742) and Thomas Simpson's *Mathematical Dissertations* (1743) and *Miscellaneous Tracts* (1757) we find the best examples of successful applications of the calculus of fluxions. As was typical in the eighteenth century, Maclaurin's and Simpson's mathematical works were motivated to a great extent by mechanics and physical astronomy. The *Principia* constituted a field of research rich in problems which demanded improvements in mechanical concepts as well as in mathematical techniques.

Basically, all the natural philosophers who decided to accept the idea of universal gravitation had to cope with three great problems: the mathematical treatment of fluids, the determination of the shape of the Earth (and the motion of the tides), and the determination of the Moon's orbit and the perturbations of the planetary motions.[1] What lay behind these problems was the necessity of extending mechanics from the study of point masses to continuous bodies. Both the study of the Earth, conceived as a rotating fluid, and that of the Moon were connected with the development of the mechanics of rigid and fluid bodies. In order to develop this programme in mechanics and astronomy, eighteenth-century mathematicians were compelled to reshape the calculus. New concepts, such as the concept of a function of many variables, and new theories, such as the theory of partial differential equations and the calculus of variations, were created in this context.

However, neither Maclaurin nor Simpson, notwithstanding the value of their researches, played an important role in the development of the eighteenth-century calculus. What they did was to extend the range of application of the original Newtonian calculus to its utmost limits. The British were unable to follow the developments of the continental calculus which was being profoundly modified.

This chapter begins with two sections devoted to Maclaurin's and Simpson's studies of the attraction of the ellipsoids. The chapter is concluded by a section dealing with Clairaut's study of the shape of the Earth. The purpose of this example is to compare the new and powerful technique of line integration with the British use of multiple integrals in studying the shape of equilibrium of a rotating fluid mass.

5.1 Maclaurin's study of the attraction of ellipsoids

The subject of the prize awarded by the French Académie des Sciences for the year 1740 was one of the main problems of Newtonian science: the flow and ebb of the tides. On 5 September 1739 De Marain, Clairaut, Nicole, Réaumur and Pitot were appointed to the committee which had the duty to select the winner. On 27 April 1740 it was announced that the prize was to be shared amongst Leonhard Euler, Daniel Bernoulli, Antoine Cavalleri, one of the last Cartesians, and Colin Maclaurin, then Professor of Mathematics at the University of Edinburgh. The choice of this subject was particularly fortunate for Maclaurin since, as appears from his correspondence (1982), he had been engaged since 1735 in research concerning the shape of the Earth. This research originated from a paper by James Stirling (1738) published in the *Philosophical Transactions* for the year 1735. Another fortunate coincidence was that Clairaut was chosen as one of the examiners for the prize essays. Clairaut himself was interested, since the Lapland expedition of 1736 in which he participated, in the problem of the Earth's shape. Indeed, he was impressed by Maclaurin's essay and was prompted to correspond with him in the years 1741–3 while preparing his *Théorie de la Figure de la Terre* (1743).[2]

We will not be concerned here with Maclaurin's prize essay (1741), which was reissued together with Euler's and Bernoulli's in the so-called Jesuit edition of the *Principia*, vol. 3 (1742),[3] but rather with the improved and augmented version of it which can be found in chapter XIV of Maclaurin's *Treatise of Fluxions* (1742).

As is well known, the mathematical study of the attraction of spheroids, of the shape of the Earth and of the tides dates back to Newton's *Principia*. Newton had assumed a fluid rotating mass as a model for the Earth and had aimed to prove that the form of relative equilibrium of such a mass is an oblate ellipsoid of revolution. Thus arose the problem of studying the attraction of points inside, on the surface and, possibly, outside a spheroid. In Proposition 90 of book I Newton found the attraction of a circular lamina on a point situated on the straight line orthogonal to the plane of the lamina and passing through the centre of the lamina. And in

Proposition 91 he showed how one could obtain the attraction of a solid
of revolution on a particle placed on the prolongation of the axis of
revolution. Another result was Corollary 3 to Proposition 91 in which it
was shown that the total gravitational force on a particle placed in a shell
bounded by two concentric, similar (i.e. with the same ellipticity) and
similarly situated ellipses is equal to zero. With this in mind one could
reduce the problem of the attraction of a particle internal to an ellipsoid to
the problem of a particle placed on the surface. From this follows also that
in the interior of the ellipsoid the attraction varies directly as the distance
from the centre. These results were used by Newton in the third book,
Propositions 18–20. Newton assumed without proof that an ellipsoid of
revolution is a form of equilibrium, and in order to find the ellipticity he
introduced the following condition: the rotating fluid ellipsoid is in
equilibrium when an equatorial and the polar infinitesimal canals balance
each other at the centre.

As can be seen, much was left to be answered. Apart from the problem
of studying fluids and their hydrostatics, there was the problem of knowing
exactly how a spheroid (to make things simpler an ellipsoid of revolution)
attracts points on its surface and external points. Only the case of points
on the axis of revolution and on the equator was solved in Newton's
Principia (1687) and Cotes's 'Logometria' (1717). Let us turn now to
chapter XIV of Maclaurin's *Treatise* (1742).[4]

After some introductory material (the geometry of the ellipse and a
recapitulation of Newton's work in this field), Maclaurin states his first
important theorem (Maclaurin (1742), article 634, pp. 524–5). Let us
consider a particle P (see fig. 2) on the surface of a homogeneous ellipsoid
of revolution and let us resolve the attraction into two components, one
perpendicular to the axis and the other one perpendicular to the plane of
the equator; then the former varies directly as the distance from the axis
and the latter as the distance from the equator. If one calculates, following
Newton and Cotes, the attractions at the pole and at the equator and uses
Maclaurin's theorem, one can calculate the variation of attraction on the
surface of the ellipsoid.

In the proof of this theorem Maclaurin confined himself to geometrical
methods. This means that he avoided the use of fluxions, fluents, algebra
or series. Indeed Maclaurin begins the chapter by proving several
propositions concerning the ellipse, and in the rest of the chapter one can
find only extremely complicated geometrical constructions. However, it is
quite evident that behind this there is the calculus.

In fact Maclaurin considered the point P on the surface as the vertex of
an infinity of infinitesimal cones in terms of which the ellipsoid was

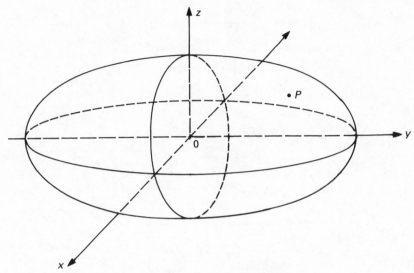

Fig. 2 Prolate ellipsoid of revolution

subdivided. He integrated once, and found the attraction of an arbitrary cone on the particle, twice, and found the attraction of an arbitrary 'slice' (a section of the ellipsoid by a plane), and a third time in order to find the total attraction. Multiple integration in spherical coordinates one could say.

Maclaurin had to use all his geometric inventiveness in order to reduce the attraction at P to known integrals. Let us consider fig. 3, which represents a section of two concentric oblate ellipsoids of revolution. $AEBQ$ and $aebq$ are two concentric ellipses of equal ellipticity; PL, parallel to the greater diameter, touches the interior ellipse at a; PK is equal and parallel to ab. The angle $f\hat{a}g$ is bisected by ab. PF and PG are drawn parallel to af and ag, respectively. Maclaurin shows (article 626) that $PH + PI = 2ai$. This equality can be used to relate the vertical component of the attraction at P to the attraction at the pole of the interior ellipsoid, which can be calculated following Newton's *Principia*, Proposition 91.

It should be noted that Maclaurin is using a differential model of the ellipsoid. That is, the volume is subdivided into infinitesimal components. Maclaurin declared in the first lines of his *Treatise* that his work was motivated by the Berkeley controversy. In fact he devoted several chapters to foundations, and here, as we know, he tried to avoid infinitesimals and moments in order to base the calculus on *ad absurdum* geometrical proofs (see chapter 3, section 3.4). We find that, quite typically of eighteenth-

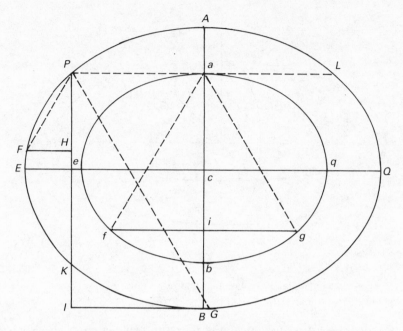

Fig. 3 Section of two concentric oblate ellipsoids of revolution

century mathematics, he forgot about that in the chapters dealing with the applications of the calculus.

Just after his theorem on the attraction at the surface Maclaurin proceeds to study the shape of equilibrium of a rotating homogeneous fluid mass. He states the following necessary conditions of equilibrium. If a fluid mass is in equilibrium, then:

(1)   the resultant force at the surface will be perpendicular to the surface;

(2)   internal surfaces similar, similarly situated and concentric to the surface of the ellipsoid will be 'level surfaces at all depths', that is the force will be perpendicular at these surfaces;

(3)   the polar and equatorial columns will balance each other at the centre;

(4)   columns drawn from an internal point to the surface will balance each other.

Let us consider conditions (1) and (2). The first one is the plumb line principle as stated by Huygens in his *Discours de la Cause de la Pesanteur* (1690): at the surface the plumb line must fall perpendicularly. The second one is more interesting. We can recognize here what we nowadays would call equipotential surfaces. At the 'level surfaces' the force is directed along the vector normal to the surface. Even though it would be excessive to see Maclaurin as a forerunner of potential theory, it cannot be denied that this extension of Huygens's principle was a promising step.

Condition (3) is Newton's principle of balancing columns, while condition (4) is understood as an extension of Newton's principle to columns meeting at an arbitrary internal point. In fact Maclaurin understands the fourth condition as follows: given an internal point draw from it an infinitesimal column to the surface, then the pressure exerted by the column on the point is independent from the direction.

Next comes the weak point in Maclaurin's treatment of forms of equilibrium. For the moment Maclaurin confined himself to homogeneous ellipsoids. He now aims to extend his analysis to cases in which the density is varying. He assumes that the density is varying according to various laws as a function of the radial distance from the centre. It was quite natural for him to take ellipsoids of revolution made up of similar and concentric shells of varying density. However, these ellipsoids, as Greenberg (1979) has pointed out, cannot be in hydrostatical equilibrium. The ellipticity of the shells has to vary as a function of density and Maclaurin did not realize this.

Another important result is the following:

Let us take two confocal ellipses and let them generate, revolving round their minor/major axis, two homogeneous ellipsoids of equal density. Let us take a point P on the prolongation of the axis of revolution (or of the plane of the equator), then the attractions of the two ellipsoids will be as their volumes.

We have here a result concerning external points: a much more difficult case than internal points. The theory of confocal ellipsoids was later extended by Legendre and Laplace. James Ivory also contributed with a paper (1809) published in the *Philosophical Transactions*. The problem was of course that of extending Maclaurin's theorem to external points not situated on the prolongation of one of the axes. As in the case of level surfaces, Maclaurin was again touching on a theme of a certain importance for potential theory.

5.2 Simpson's study of the attraction of ellipsoids

Just one year after the publication of Maclaurin's *Treatise of Fluxions* (1742), Simpson included two essays dealing with the attraction of ellipsoids in his *Mathematical Dissertations on a Variety of Physical and*

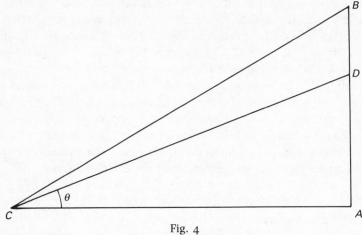

Fig. 4

Analytical Subjects (1743).[5] He adopted Maclaurin's technique of slicing up the ellipsoid into infinitesimal 'wedges', but employed the calculus to perform the necessary multiple integrations. Thomas Simpson begins from this simple lemma (see fig. 4):

Supposing *AC* perpendicular to *AB*, and that a Corpuscle at *C* is attracted towards every Point or Particle of the line *AB*, by Forces in the reciprocal duplicate Ratio of the Distances; to determine the Ratio of the whole Force whereby the Corpuscle is urged in the direction *CA*. (Simpson (1750c), II, p. 445)

If we now consider a variable point *D*, and we put $DA = x$ and $CA = a$, we will have that the horizontal component of the force at *D* ($dF \cos \theta$) will be proportional to $AC/CD^3 = a/(a^2 + x^2)^{\frac{3}{2}}$:

Therefore

$$\frac{a\dot{x}}{(a^2 + x^2)^{\frac{3}{2}}}.$$

is the Fluxion of the whole Force, whose Fluent [...] is =

$$\frac{x}{a(a^2 + x^2)^{\frac{1}{2}}}.$$

(Simpson (1750c), II, p. 446)

Integrating, one finds that the total force has a horizontal component proportional to $BA/(CA \cdot CB)$.

A second lemma allows one to calculate the attraction of an elementary 'wedge' and reads as follows (see fig. 5):

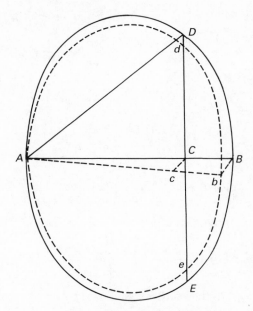

Fig. 5 From Simpson (1750c), II, p. 454

Supposing *ADBbA* to be a Cuneus of uniformly dense Matter, compriz'd by two equal and similar eliptic Planes *ADBEA* and *AdbeA*, inclin'd to each other, at the common vertex *A*, of either their first or second Axes, in an indefinitely small angle *BÂb*; to determine the Attraction thereof at the Point *A*, supposing the Force of each Particle of Matter to be as the Square of the Distance inversely. (Simpson (1750c), II, p. 453)

Let $AB = a$, $BC = x$, $CD = y$, $\sin (B\hat{A}b) = d$ and $y^2 = fx - x^2 - gx^2$. Then we have $Cc = d(a - x)$.
But $AD = ((a-x)^2 + fx - x^2 - gx^2)^{\frac{1}{2}}$, and, for the first lemma,

$$\frac{DE \times Cc}{AC \times AD}$$

expresses 'the Attraction of the Particles in the indefinitely narrow Rectangle *DE·Cc*', which will be equal to

$$\frac{2d(fx - x^2 - gx^2)^{\frac{1}{2}}}{((a-x)^2 + fx - x^2 - gx^2)^{\frac{1}{2}}}$$

The 'fluxion of the force to be found' will then be:

$$\frac{2d(fx - x^2 - gx^2)^{\frac{1}{2}}\dot{x}}{((a-x)^2 + fx - x^2 - gx^2)^{\frac{1}{2}}}$$

(Simpson (1750c), II, p. 455)

But $fx - x^2 - gx^2 = 0$, when $x = f/1 + g = a$, or $x = 0$. Therefore, substituting:

$$\frac{2d((1+g)ax - (1+g)x^2)^{\frac{1}{2}}\dot{x}}{((a-x)^2 + (1+g)ax - (1+g)x^2)^{\frac{1}{2}}} =$$

$$\frac{2d(1+g)^{\frac{1}{2}}x^{\frac{1}{2}}\dot{x}}{(a+gx)^{\frac{1}{2}}} =$$

$$\frac{2d(1+g)^{\frac{1}{2}}x^{\frac{1}{2}}}{a^{\frac{1}{2}}}(1+gx/a)^{-\frac{1}{2}}\dot{x} =$$

$$2d(1+g)^{\frac{1}{2}}x^{\frac{1}{2}}a^{-\frac{1}{2}}\left(1 - \frac{1}{2}\frac{gx}{a} + \frac{3}{2\cdot4}\left(\frac{gx}{a}\right)^2 - \frac{3\cdot5}{2\cdot4\cdot6}\left(\frac{gx}{a}\right)^3 + ...\right)\dot{x}.$$

Therefore, integrating from $x = 0$ to $x = a$:

$$2d(1+g)^{\frac{1}{2}}a\left(\frac{2}{3} - \frac{2}{5}\frac{g}{2} + \frac{2}{7}\frac{3}{2\cdot4}g^2 - \frac{2}{9}\frac{3\cdot5}{2\cdot4\cdot6}g^3 + ...\right).$$

$(1+g)^{\frac{1}{2}}a = (1+g)^{-\frac{1}{2}}f$, therefore expanding $(1+g)^{-\frac{1}{2}}$ in powers of g and multiplying the two series:

$$2df\left(\frac{2}{3} - \frac{2\cdot4}{3\cdot5}g + \frac{2\cdot4\cdot6}{3\cdot5\cdot7}g^2 - ...\right).$$

Now Simpson can study the attraction at the surface of the ellipsoid. As can be seen in fig. 6, he follows a technique very similar to Maclaurin's. Simpson considers two sections of the ellipsoids generated by two planes passing through Q and perpendicular to the plane of the meridian. It is easy to show that these two sections are two ellipses. Supposing them to revolve around an 'indefinitely little' angle, they will determine two elementary 'wedges'. Applying the preceding result Simpson can determine the attraction exerted by the two elliptical 'wedges', in the directions QH and Qh, respectively. The integration needed to find the attraction of the whole mass is performed as before by power series.[6]

Reading Maclaurin and Simpson is interesting for us because it is possible to refute some widespread opinions about the fluxional school. It is often maintained that eighteenth-century British mathematicians were far behind the continentals because (1) they compelled themselves not to use infinitesimal techniques, and (2) they rejected 'analysis' in favour of 'geometry'. As far as the use of infinitesimals is concerned, we can see how both Maclaurin and Simpson had no problems in employing the infinitesimals, even though they spurned them at the foundational level. It is to be noted that, generally, they introduce the infinitesimal in a non-

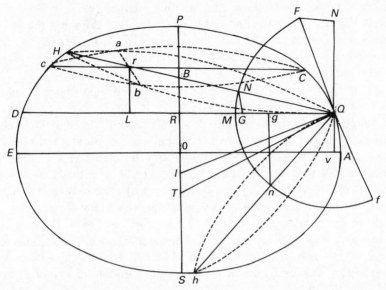

Fig. 6 From Simpson (1750c), II, p. 456

kinematic way. This sharply conflicts with their foundations in terms of
kinematic concepts. Furthermore, we can surely say that the use of
'analysis' characterizes Simpson's *Mathematical Dissertations* (1743) to an
extent which is sufficient to render the usual image of the fluxional
calculus inappropriate to his work. The case of Maclaurin is more complex.
His research on ellipsoids is included in the first book of the *Treatise* (1742),
in which he did not employ the calculus. This was an epistemological
choice. The integrations necessary for proceeding with the proofs
analytically (we have seen some instances in Simpson) were easy for any
good mathematician at that period. In fact, in the second book Maclaurin
provided the integrals required for calculating the attraction at the pole
and at the equator of an ellipsoid (Maclaurin (1742), pp. 722–6). It was
easy for him to translate his 'geometry of fluxions' into a 'calculus of
fluxions', but the methodological programme of the *Treatise* (1742)
consisted in showing that the calculus, even in its most advanced parts,
was based on geometry.

It is certain that Maclaurin and Simpson greatly extended the study of
attraction of ellipsoids: they calculated the attraction of any internal
particle of an homogeneous ellipsoid of revolution. Maclaurin knew that
the potentials of confocal ellipsoids of equal density at points situated on
the prolongation of the axes are related directly to their volumes. In
treating fluid equilibrium Maclaurin made use of an interesting concept

which anticipates somehow that of equipotential surfaces. As far as I know, Maclaurin was the first one to study confocal ellipsoids, the first one who gave an exact calculation of the variation of attraction on the surface of an ellipsoid of revolution, and the first one who introduced the concept of level surfaces.

Maclaurin's and Simpson's influences on British authors cannot be overestimated. After more than sixty years, in the chapter concerned with the shape of the Earth of John Robison's *Elements of Mechanical Philosophy* (1804) one could find Maclaurin and Simpson and not Laplace. But, how much influence was exerted by them on the continentals? Maclaurin was certainly very well known. His prize essay (1741) was available in Latin, while two copies of Maclaurin's *Treatise* (1742) were sent to France, one to the Académie and the other one as a complimentary copy to Clairaut; it was translated into French in 1749. As we will see below, Clairaut in his *Théorie de la Figure de la Terre* (1743) declared his debt to Maclaurin: he adopted level surfaces, calling them *surfaces courbes de niveau* and even adopted Maclaurin's geometrical proof of the variation of attraction at the surface of the ellipsoid. Also, d'Alembert, Lagrange and Laplace tributed to Maclaurin the merit of having introduced the concept of level surfaces and praised his geometrical methods. However, notwithstanding this unanimous appreciation, Maclaurin was soon superseded on the continent and became of little use for the researcher. Indeed already in 1743 with Clairaut's *Théorie de la Figure de la Terre* we find mathematical methods that render Maclaurin's and Simpson's works obsolete. That is, Clairaut was not basing his mathematical analysis only on multiple integration, but he used line integrals. In fact, Clairaut considered curvilinear infinitesimal canals internal to the fluid and integrated the force along these canals, finding what we nowadays would call the work. In John Greenberg's Ph.D. thesis (1979) there is a long section devoted to Clairaut's study of the shape of the Earth. Quite appropriately Greenberg aims at showing that it is in the context of applied mechanics and hydrostatics that Clairaut devised a new mathematical tool, line integration.

In the 1730s and 1740s the continental mathematicians deeply transformed the calculus. It is in these years that we find some use of line integrals (especially in Clairaut), the first steps towards the calculus of variations (culminating with Euler's *Methodus inveniendi* in 1744), and later on partial differential equations (d'Alembert's vibrating string, for instance). These tools were for some reason outside the scope of the fluxionists. The British mathematicians, very often gifted mathematicians, were unable to develop the calculus as successfully as their continental colleagues. In eighteenth-century British calculus the absence of improvements in a multivariate calculus is especially remarkable. It is the multivariate calculus (line integrals, partial differential equations and the calculus of variations) that allowed the continentals, from Clairaut to d'Alembert, from Lagrange to Laplace, to solve the basic problems of Newtonian science, such as the determination of the Earth's shape.

Maclaurin and Simpson are therefore typical Newtonians. In the case considered in this chapter they successfully applied the geometrical tools of the *Principia* (1687), the integrals of 'De quadratura' (1704c) and Cotes's 'Logometria' (1717) to tackle the attraction of ellipsoids. Their skill in going so far with only the help of geometrical inventiveness catches the applause; in the same time the absence of multivariate calculus techniques renders them old fashioned mathematicians who could not participate in a period of the history of mathematics in which the calculus underwent a deep change.

5.3 Remarks on the use of partial differentials in Clairaut's *Théorie de la Figure de la Terre*

Clairaut's *Théorie de la Figure de la Terre* was published in 1743. Its second part was explicitly based on Maclaurin:[7]

I have decided to treat in detail the figure of homogeneous spheroids, and to abandon my method, as far as these spheroids, in order to follow the one which Maclaurin gives in his excellent Traité des Fluxions. This method seems to me so beautiful and profound, that I hope its insertion here will please my readers. (Clairaut (1743), p. 158)

Nevertheless, Clairaut's work rendered obsolete that of both Simpson and Maclaurin since he employed mathematical concepts and techniques unknown to the fluxionists. We can see this, for instance, in the fourth chapter 'Maniere générale de faire usage du principe de l'équilibre des canaux de figure quelconque' (Clairaut (1743), p. 33). The 'principe' had already been given in the first chapter and states that a fluid mass will be in equilibrium when any infinitesimal canal either traversing the mass or returning to itself is in equilibrium. In the case of spheroids Clairaut thinks that the condition of equilibrium can be expressed as follows:[8]

In order that a fluid rotating spheroid, for which the law of gravity is given, can maintain a constant shape, it will be sufficient that an arbitrary channel, re-entering to itself and placed in a meridian plane of this spheroid, is always in equilibrium, if one takes into account only the force of gravity and not the centrifugal force. (Clairaut (1743), pp. 14–15)

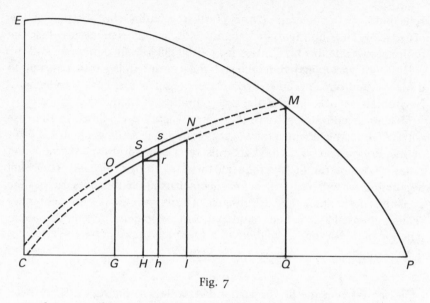

Fig. 7

He then considers a section CEP (see fig. 7) of the spheroid, where P is
the pole and C is the centre. Let ON be an infinitesimal canal lying on the
plane of the meridian CEP. If the condition of equilibrium is satisfied, then
the pressure exerted by ON is equal to the pressure exerted by any other
canal terminated by O and N. Let $CH = x$, $HS = y$, $Sr = dx$ and $sr = dy$. Let
the 'force of gravity perpendicular to CP' be P, and let Q be the 'force of
gravity parallel to CP' (Clairaut (1743), p. 35). Then '$Pdy + Qdx$ will be the
total effort of the small Cylinder Ss due to these two forces'. If the fluid is
in equilibrium, the value of

$$\int Pdy + Qdx$$

will not depend on the path of integration.

In treating the integrability of $Pdy + Qdx$, Clairaut says:[9]

If one wished to use this quantity, in order to determine in finite terms the value
of the weight of the channel ON, under the assumption that the curvature of this
channel is given, one might begin eliminating y and dy from $Pdy + Qdx$; this
differential would now have only x and dx, and it would be possible to integrate it
after having completed the integral, that is after having found the necessary
constant which would render the weight equal to zero; taking x equal to CG, and
then $x = CI$, one would be able to determine the total weight of the channel ON.
But since the equilibrium of the fluid requires that the weight of ON does not
depend on the curvature of OSN, that is to say on particular values of y and x, it
follows that $Pdy + Qdx$ should be integrated without knowing the value of x, that
is to say *it is necessary that $Pdy + Qdx$ is a complete differential (différentielle complette)
as a condition for having equilibrium in the fluid.* (Clairaut (1743), pp. 36–7)

Here we have a very good example of how continental mathematicians were beginning to apply a multivariate calculus to mechanics. Since the form of the canal *ON* is arbitrary, we cannot eliminate a variable: it is essential for Clairaut that both x and y should be considered as independent variables. The mechanical model of the 'rentrant' canals, much more general than that of the 'elementary wedges' employed by Maclaurin and Simpson, was not mathematically approachable without an understanding of a *function of many variables* and a *partial differential*.

Clairaut was able to use in (1743) some results he had published in 1740. He established that $Pdy + Qdx$ is a '*différentielle complette*' when $dP/dx = dQ/dy$ (to the best of my knowledge nobody questioned the generality of this result during the eighteenth century). For '*différentielle complette*' Clairaut meant:[10]

a quantity whose integral is a function of x and y. $ydx + xdy$, $(ydx + xdy)/2\sqrt{(aa + xy)}$ are complete differentials, because their integrals are xy, $\sqrt{(a^2 + xy)}$. (Clairaut (1743), p. 37n)

While the symbol dP/dx (i.e. our $\partial P/\partial x$) is to represent[11]

the differential (*différentielle*) of the function P, taken supposing only x as variable. (Clairaut (1743), p. 38n)

Like Clairaut, both d'Alembert and Euler, who made fundamental contributions to fluid mechanics, were to employ a calculus of functions of many variables. It is at this level that an important distinction can be made between the fluxional and the differential calculus. It seems to me that the case of the study of ellipsoids shows that the fluxionists were able to employ analytical techniques in which the notion of infinitesimal and the principle of cancellation of higher order differentials occurred; i.e. they were prepared to use in the applications a calculus very similar to that of the early Leibnizians. However, they did not accept the mathematical techniques developed on the continent in the 1730s which transformed the calculus of differentials into a calculus of multivariate functions.

6

THE ANALYTIC ART
(1755–85)

THE PICTURE of the fluxional calculus that one can derive from the textbooks written for teaching purposes is not completely representative of the level and methodology of the research carried out by British mathematicians. It was not the kinematical method of fluxions but rather another tool, labelled vaguely as 'analytics', which was adopted by some of the most active fluxionists. Among them the most notable were: Thomas Simpson (1710–61), John Landen (1719–90) and Edward Waring (1736–98). 'Analytics' had a different meaning for each of the three: however, a general idea can be inferred from their works.

'Analytics' was a calculus not immediately interpretable in terms of geometrical or kinematical concepts. The use of symbolism was free from the constraints of interpretability: imaginary numbers, logarithms of negatives and so on were considered admissible. This freedom was not usual in eighteenth-century Great Britain.[1] The limits imposed by the kinematical conception of mathematical quantities were ignored because it was thought that the progress of the fluxional calculus was directly connected with the development of analytical methods. For instance, a formal use of series was considered fundamental: the analytical fluxionists were hoping to attain results by working algebraically on infinite series.

The research on the continent was thought of very highly by the analytical fluxionists, and a genuine attempt was made to import continental methods. This early attempt of reforming the calculus was largely a failure. As we have seen in the previous chapter, the fluxionists continued to understand a fluent x as a single variable function $x(t)$. 'Taking the fluxion' was understood as differentiation as a function of t; in consequence only total differentials were handled. On the contrary, the continental calculus in the middle of the eighteenth century was centred upon objects such as partial differential equations, line integrals and variational techniques which could not be translated into the fluxionists' conceptual framework.

The analytical researches of Simpson, Landen and Waring passed almost unnoticed by their fellow countrymen and by the continentals. It is easy to see why this happened if one compares the production of Euler or Lagrange with the modest contributions of the British. However, Simpson's papers on series (1759b) and 'isoperimetrical' problems (1756a) and (1759a), Landen's results on elliptic integrals (1772) and (1775), and Waring's attempt to develop a multivariate calculus (1776), deserved to receive more attention in Britain: it was only by following their steps that the fluxionists could continue to be competitive, as they had been in the first half of the century with Stirling, Taylor, Cotes and Maclaurin.

6.1 Simpson's methodology

Thomas Simpson was one of the fluxionists who practised the analytical method. In his *Doctrine and Application of Fluxions* (1750c) he accepted Maclaurin's kinematical foundation of the calculus. But in the same work we find a hint of another approach:

But, after all, the *Differential Method* [i.e. Newton's *methodus differentialis*] has one Advantage above that of Fluxions, which is, we are not there obliged to introduce the Properties of Motion. Since we reason upon the Increments themselves, and not upon the Manner in which they may be generated.

It has been hinted above, that, though the Increments of Quantities are not, *strictly*, as the Fluxions, yet from them the Ratio of the Fluxions may be deduced; and it appears that the smaller those Increments are taken, the nearer their Ratio will approach that of Fluxions. Therefore, if we can, by any Means, find the Ratio to which the said Increments, by conceiving them less and less, do perpetually converge, and which they may approach, before they vanish, nearer than any assignable Difference, that Ratio (called, hereafter, for Distinction Sake, the *Ratio limiting that of the Increments*) will be, *strictly*, that of the Fluxions. (Simpson (1750c), p. 152)

Stating the dependence of the method of fluxions on the non-kinematical method of finite differences was not that new; as we already know, Taylor had maintained this position. However, Simpson's non-kinematical approach to the fluxional calculus appeared provocative after Maclaurin. He had no intention of confining himself within the limited field of geometrically interpretable mathematics. Simpson had the 'foreign mathematicians' as models:

I have chiefly adhered to the *analytic method of Investigation*, as being the most direct and extensive, and best adapted to these abstruse kinds of speculations.

Where a *geometrical demonstration* could be introduced, and seemed preferable, I have given one: but tho' a problem, sometimes, by this last method, acquires a degree of perspicuity and elegance, not easy to be arrived at any other way, yet I cannot be of the opinion of Those who affect to shew a dislike to every thing performed by means of *symbols* and an *algebraical Process*; since, so far is the *synthetic method* from having the advantage in all cases, that there are innumerable enquiries into nature, as well as in abstracted science, where it cannot be at all applied, to any purpose. Sir Isaac Newton himself (who perhaps extended it as far as any man could) has even in the most simple case of the lunar orbit (Princip. B. 3. prop. 28) been obliged to call in the assistance of *algebra*; which he has also done, in treating of the motion of bodies in resisting mediums, and in various other places. And it appears clear to me, that, it is by a diligent cultivation of the *Modern Analysis*, that *Foreign Mathematicians* have, of late, been able to push their *Researches* farther, in many particulars, than Sir Isaac Newton and his *Followers* here, have done: tho' it must be allowed, on the other hand, that the same *Neatness*, and *Accuracy* of *Demonstration*, is not every-where to be found in those *Authors*, owing in some measure, perhaps, to too great a disregard for the *Geometry of the Ancients*. (Simpson (1757), Preface)

Even though Simpson declared his appreciation of the analytical methods adopted by the continentals, he found it difficult to follow the recent developments of the calculus. For instance, in his papers (1756a) and (1759a) on isoperimetrical problems, he did not cite Euler's *Methodus Inveniendi Lineas Curvas* (1744). Simpson's style is so clumsy that one is led to think that he did not even know Euler's work. His *General Proposition* reads as follows:

Let Q, R, S, T, &c. represent any variable quantities, expressed in terms of x and y (with given coefficients), and let q, r, s, t, &c. denote as many other quantities, expressed in terms of $\dot{x}$ and $\dot{y}$; It is proposed to find an equation for the relation of x and y, so that the fluent of $Qq + Rr + Ss + Tt$, &c. corresponding to a given value of x (or y), may be a maximum or minimum. (Simpson (1759a), p. 624)

Here one cannot find either the notation or the concept of function which played so relevant a role in Euler's treatment of variational problems. However, in functional notation Simpson expresses the problem of finding the extremals of $v[y(x)] = \int_{x_1}^{x_2} F(x, y(x), y'(x))dx$, and arrives at a *General Rule*, which can be understood as a fluxional version of Euler's equation:

$$F_y - \frac{d}{dx} F_{y'} = 0.$$

Furthermore, he solved problems with side conditions with the method of 'Lagrange' multipliers. These papers represent a step towards a general treatment of 'isoperimetrical problems', which had received scarce attention in Great Britain. Only special cases, such as the brachistocrone or the solid of least resistance, had previously been considered by British mathematicians.

In addition, Simpson's work on series can be considered promising: for instance, in (1759b) he worked on series with complex coefficients. However, we cannot mention any important result achieved by Simpson in this field. But certainly the level of his researches on series is above the average. Sometimes in the history of mathematics, especially when a process of reform is taking place, new methodological approaches are as important as new results. And, from this point of view, it is important to notice the formal use of series, which characterizes Simpson's papers (1753) and (1759b) as well as the contributions on series in his books (1740) and (1743). In these works series are treated as algebraical expressions, independently from their 'numerical' interpretation (i.e. complex numbers as well as divergent series are handled freely in the algorithmic procedures). As we will see in the next chapters, the formal use of series was to play an important role in the reform of British calculus. In fact, the early nineteenth-century reformers (e.g. Robert Woodhouse and the Analytical Society's fellows) adopted the 'Lagrangian foundation', developed by Lagrange in his *Théorie des Fonctions Analytiques* (1797), according to which all the theorems of the calculus should be reduced in terms of algebraical manipulations with series. It is therefore interesting to find that *some* Newtonians worked with series developments following a methodology which was compatible with the Lagrangian foundation of the calculus.

6.2 Landen's *Residual Analysis*

John Landen (1719–90) was a land surveyor who lived in Northampton-shire.[2] He was a friend and correspondent of Simpson. In his letters he showed a great interest in Simpson's analytical research.[3] Like Edward Waring, John Colson, Francis Maseres and Robert Smith, Thomas Simpson was a subscriber to *The Residual Analysis* (1764), Landen's chief work.

In *The Residual Analysis* Landen tried to reformulate the principles of the calculus on a completely new basis. His purpose was to base the calculus upon a purely algebraical procedure in order to avoid both Newton's fluxions and Leibniz's differentials:

In the application of the *Residual Analysis*, a geometrical or physical problem is naturally reduced to another purely algebraical; and the solution is then readily obtained, without any supposition of motion, and without considering quantities as composed of infinitely small particles. (Landen (1758), p. 5)

Landen did not oppose the calculus of fluxions on the grounds that it was faulty, but because it was just a limited case of a more general theory:

If any thing can be said by way of objection to the fluxionary method, it is, that the new principles on which it is founded, though accurate, are not the genuine principles of Analytics, for the improvement of which, those principles were borrowed from the doctrine of motion: And that, although such borrowed principles may enable us to give very concise solutions to certain problems, yet perhaps we must not expect to bring the Analytic art to its utmost perfection, otherwise than by proceeding upon its own proper principles. What weight there may be in such objection, I shall not take upon me to determine. Yet I must confess, that, how natural soever it may be, in the resolution of certain problems relating to geometrical magnitudes, to consider such magnitudes as generated by motion, it seems to me not natural to bring motion into consideration in resolving problems purely algebraical: Nor does it seem natural, in resolving problems concerning the motion of bodies, to superinduce imaginary motions, and therewith bring into consideration the *velocity* of *time*, the *velocity* of a *velocity*, &c. nor yet does it appear more natural, in the resolution of other problems, to make use of the fluxionary method, when (as is most commonly the case in that doctrine) the fluxions introduced into the process can only in a figurative sense be said to be the velocities of increase of the quantities called their fluents; such figurative expression being not the natural language of Analytics, but frequently, instead of conveying clear and distinct ideas, is confusedly employed in treating of quantities as generated by motion, which in reality cannot be conceived to be so generated (such quantities are weight, density, &c.) (Landen (1764), pp. iv–v)

In short, Landen's idea consisted in using the formula:

$$\frac{x^{m/n} - v^{m/n}}{x - v} = x^{m/n-1} \frac{\sum\limits_{z=0}^{m-1} (v/x)^z}{\sum\limits_{w=0}^{n-1} (v/x)^{\frac{wm}{n}}}.$$

In order to differentiate $x^{\frac{4}{3}}$, Landen wrote:

$$\frac{x^{\frac{4}{3}} - v^{\frac{4}{3}}}{x - v} = x^{\frac{4}{3}-1} \frac{\sum\limits_{z=0}^{3} (v/x)^z}{\sum\limits_{w=0}^{2} (v/x)^{\frac{w4}{3}}}.$$

When $x = v$, he obtained the 'special value' $\frac{4}{3}x^{\frac{1}{3}}$.

Landen's procedure is open to several objections. First of all, this algebraical foundation of the calculus works only with rational functions. Let us consider, for instance, x^α, α irrational: in this case Landen is compelled to introduce a limit process.

If we take $\alpha = \sqrt{2}$ we will have, in Landen's notation:

$$\frac{x^\alpha - v^\alpha}{x - v} = x^{\alpha-1} \frac{1 + (v/x) + (v/x)^2 + \ldots(1414\text{ times})}{1 + (v/x)^\alpha + (v/x)^{2\alpha} + \ldots(1000\text{ times})}.$$

Or, 'more nearly',

$$\frac{x^\alpha - v^\alpha}{x - v} = x^{\alpha-1} \frac{1 + (v/x) + (v/x)^2 + \ldots(14\,142\text{ times})}{1 + (v/x)^\alpha + (v/x)^{2\alpha} + \ldots(10\,000\text{ times})}.$$

Therefore, in order to calculate the 'special value', Landen is led to calculate what he calls the 'ultimate value' (Landen (1764), p. 8) of

$$\frac{1 + 1 + 1 + \ldots(14\,142\text{ etc.})}{1 + 1 + 1 + \ldots(10\,000\text{ etc.})}.$$

A second objection is that the fundamental equation of the *Residual Analysis* is valid only if $x \neq v$, in fact the left term of Landen's equation when $x = v$ is equal to $0/0$. In modern terms, Landen tries to assign a definite value for $x = r$ to $f(x)/g(x)$, when $f(r) = g(r) = 0$. He thought he had found $R(x) = f(x)/g(x)$ such as $R(r)$ is algebraically defined. In this case a conception of algebraical equality comes into play which was often adopted in the eighteenth century. An algebraical equation was to be taken as universally true: if on one side of the equation a substitution was not possible, this depended upon the particular form of $f(x)/g(x)$. The other side expressed the 'general value of such fractional quantity in terms which shall not vanish when x is equal to r' (Landen (1758), p. 8), and therefore the 'special value' could be attained.

Even though the foundation of the calculus proposed in *The Residual Analysis* is not successful, Landen's work is remarkable from several points of view. It was a systematic attempt to de-geometrize the calculus and to work only with algebra. Algebra, or 'analytics', dealt with universally true equalities which could be interpreted according to the kinematical model of fluent quantities. Landen's methodology somewhat resembles that developed and explicated by the French Lagrangian school and by the analytical school of Cambridge in the 1810s; this methodology would play a relevant role in the period spanning the eighteenth and nineteenth centuries. Landen's terminology broke with the fluxional tradition: in *The Residual Analysis* one finds a definition, probably taken from Euler, of a 'function' as 'an algebraic expression composed, in any manner, of any power or powers of any variable quantity, with any invariable coefficients' (Landen (1764), p. 2). In fact, Landen does not conceive the calculus as dealing with variables, their velocities, finite increases, and so on, but with 'algebraic expressions' and 'functions'. As in the case of Simpson, we can see here the beginnings – unfulfilled – of an algebraic approach to the calculus.

It is interesting to find in chapter IV of *The Residual Analysis* three theorems and a corollary on the continuity of certain classes of functions. The first one reads as follows:

Suppose E to be an algebraic expression composed of x and other quantities; and suppose, that, how near soever x be taken to some certain quantity g, E is positive when x is less than g, and negative when x is greater than g; or positive when x is greater than g, and negative when x is less than g; then shall E, or its reciprocal, be $= 0$ when x is $= g$. (Landen (1764), p. 46)

So Landen implicitly assumes that the only discontinuities are vertical asymptotes. If there is in g such a discontinuity, the reciprocal of the 'algebraic expression' will have a zero in g. Continuity was generally assumed to be a universal property of functions by Landen's contemporaries, so it is no surprise to find in the proof of the preceding proposition that:

If the value of E approaches to 0 when x is taken nearer and nearer to g, it is evident that E will be $= 0$ when x is $= g$. (Landen (1764), pp. 46–7)

The second proposition depends upon the first. It states that if $Q(x)$ is positive for $x = g$, then $Q(x)$ will be positive when 'x is greater or less than g between certain limits'. The third one says that if $Q(g) > 0$ and m is an odd number or a 'fraction whose numerator and denominator are both odd numbers', then there is a neighbourhood of g such that $Q(x) \cdot (x-g)^m$ 'will be negative when x is less than g, and positive when x is greater than g' (Landen (1764), pp. 47–8). These, of course, are very simple propositions in which restrictive suppositions on continuity are implicitly made. However, the style is completely new in the history of the fluxional calculus. In order to justify the calculation of the derivative (the 'special value'), Landen does not refer to the properties of motion but studies the properties of continuity of algebraical expressions. In fact, in *The Residual Analysis* the propositions of chapter IV are employed to prove the correctness of the substitution of a variable which allows the passage from the 'general value' to the limiting 'special value'.

The algebraical technique developed in *The Residual Analysis* was not employed by Landen in his other researches. For instance, in his works on integration he explicitly employs a method of limits. His most famous results are the so-called Landen's transformations for elliptic integrals, which were published in the *Philosophical Transactions* for the years 1771 and 1775. Here Landen employed both the notation and the terminology of the fluxional calculus; his algebraic approach to the calculus was not even hinted at. It is in this context that Landen proved that an arc of an hyperbola can be expressed by means of two elliptic arcs.

Studying the difference between the arc and the tangent of an hyperbola and a similar expression for the ellipse, Landen arrives at the following fluxional equation:

$$\frac{(\tfrac{1}{2})m^{-\frac{1}{2}}n^2y^{-\frac{1}{2}}\dot{y}}{(n^2+2fy-y^2)^{\frac{1}{2}}}+\frac{(\tfrac{1}{2})m^{-\frac{1}{2}}n^2z^{-\frac{1}{2}}\dot{z}}{(n^2-2fz-z^2)^{\frac{1}{2}}}=0.$$

(Landen (1772), p. 305) which holds when the variables y and z are linked by the relation

$$(*)\; mn^2-n^2y-n^2z-myz=0.$$

It would be excessive to attribute to Landen a theory of elliptic integrals. His problem was not that of integrating functions of the form:

$$((1-x^2)(1-q^2x^2))^{\frac{1}{2}},$$

and he never expressed Landen's transformations in the form known today. In fact the relation (*) is just a passage in the calculation and is not given particular attention; what is interesting for Landen is the geometric relation between the arc of the hyperbola and the arc of the ellipse. However, Landen's papers suggested to Lagrange and Legendre important generalizations which belong to the theory of elliptic integrals. Again, another fluxionist was shaping fruitful trends of research in the development of 'analytics'.[4]

6.3 Waring's work on partial fluxional equations

The case of Edward Waring (1736–98) is indicative of the conceptual difficulties that fluxionists could come across in promoting the analytical approach to the calculus.[5] His research in the field of the calculus deserves our attention not because of its value, but also because it constitutes a failure by one of the most talented British mathematicians. Waring was educated at Cambridge and was elected Lucasian Professor, in succession to John Colson, in 1761. It seems that he did not lecture; he spent his time in isolation, researching into mathematics and anatomy. His best work was on algebraic equations (see Waring (1762)), but he also produced a sizeable volume entitled *Meditationes Analyticae* (1776; 2nd edn. 1785) on fluents, fluxional equations, series and finite differences. In the preface he showed a good knowledge of the works of Clairaut, d'Alembert and Euler.[6] He lamented the fact that in Great Britain mathematics was cultivated with less interest than on the continent, and clearly desired to be considered as highly as the great names in continental mathematics.

There is no doubt that Waring was reading continental mathematics at a level never reached by any other British mathematician, but what he accomplished was very often a confused and sometimes misleading translation of the works of the continentals into a different conceptual framework. The case of partial differential equation is revealing.

The first proposition to be considered is Theorem II:[7]

Given a quantity A, in which two variable quantities x and y occur; let its fluxion be $\dot{A} = a\dot{x} + b\dot{y}$; let us determine the fluxions of the quantities a and b, which will be respectively $\dot{a} = \alpha\dot{x} + \beta\dot{y}$, $\dot{b} = \pi\dot{x} + \rho\dot{y}$, where a, b, α, β, π, ρ, are functions of the characters x and y; therefore it will be $\pi = \beta$, that is π will be the same quantity as β.

Corollary. From this it follows that it can be found, given a fluxion $(\dot{A} = a\dot{x} + b\dot{y})$ which involves two variable quantities, if its fluent can be expressed, or not. In fact, let us determine the fluxions of the quantities a and b, and if $\dot{a} = \alpha\dot{x} + \beta\dot{y}$, and $\dot{b} = \pi\dot{x} + \rho\dot{y}$, and $\pi = \beta$, then the fluent can be expressed; if this does not hold the fluent cannot be expressed. (Waring (1776), 2nd edn., pp. 33–4)

At first sight, this theorem could be understood as dealing with multivariate functions; but here Waring is merely restricting his analysis to ordinary exact differential equations. The corollary expresses the conditions of integrability. The quantity A is understood by Waring as $A(x(t), y(t))$ and its fluxion is obtained by implicit differentiation and should be written as

$$\frac{\mathrm{d}A}{\mathrm{d}t} = \frac{\partial A}{\partial x}\frac{\mathrm{d}x}{\mathrm{d}t} + \frac{\partial A}{\partial y}\frac{\mathrm{d}y}{\mathrm{d}t}.$$

Indeed, Waring deals only with $F(x(t), y(t) \ldots z(t))$ and always conceives the process of differentiation as $\mathrm{d}/\mathrm{d}t$; in consequence he can only obtain ordinary differential equations. The restriction imposed by the corollary to Theorem II indicates that Waring did not grasp the concept of a multivariate function. This proposition was already known to Clairaut: but we have seen that in *Théorie de la Figure de la Terre* (1743) the mechanical model to which exact differentials were applied imposed the independence of both the variables x and y.

However, at the end of chapter III of the *Meditationes Analyticae* (1776) Waring presents some partial differential equations taken from Euler's *Istitutiones Calculi Integralis* (1768–70): here Euler's bracketed notation for partial derivatives is adopted. We find, for instance, the vibrating string equation expressed as[8]

$$\left(\frac{\ddot{V}}{\dot{y}^2}\right) = a^2\left(\frac{\ddot{V}}{\dot{x}^2}\right).$$

All the partial differential equations considered by Waring with their solutions are confusedly copied from Euler: it is impossible to follow

Waring's procedure because he contracts into three lines what Euler does in five pages of well ordered explication. He picks up things haphazardly from the *Istitutiones*, jumping from one chapter to another. Waring never acknowledges his indebtedness to Euler, even though not only Euler's theorems but also bits of his Latin text were copied. It is difficult for the reader to understand the role played by partial differential equations within the *Meditationes Analyticae* (1776). The traditional definition of fluent and fluxion given in the first chapter, followed by Theorem II, allows no possibility of understanding, for instance, what the vibrating string equation might mean. Furthermore, Waring never hints at the mechanical applications of the calculus and does not deal at all with the calculus of variations. It is important to stress that Waring was a very good mathematician and that nobody else in Great Britain had tried to approach Euler's mathematics. The difficulties he found in reading the well ordered and didactic exposition of partial differential equations in the third volume of the *Istitutiones Calculi Integralis* is a measure of the gap between the fluxional and the 'Eulerian' conceptualization of the calculus.

PART III

THE REFORM

7

SCOTLAND
(1785–1809)

THE NEXT three chapters are devoted to the attempts at reforming the calculus which were made in the late eighteenth century and early nineteenth century. The first location to which our attention is drawn is Scotland, where it was more likely that there were contacts with the continent. A geometric tradition was alive in Glasgow and Edinburgh, but this did not hinder several Scottish mathematicians from appreciating the importance of the 'analysis' elaborated by the continentals. John Playfair played a significant role in promoting the study of analytical mathematics. In his lectures on physical astronomy in Edinburgh he certainly introduced a taste of the new techniques of Laplace. Perhaps more influential were his reviews and historical accounts concerning the development of mathematics in which he was extremely critical of the fluxional tradition. His call for a change was taken up by two young Scots, Wallace and Ivory. Their researches on elliptic integrals, motivated by Landen (1775), led them to appreciate Legendre's work. I conclude the chapter with a few notes on a quite obscure outsider, William Spence, who adopted in the early nineteenth century the differential notation and, more importantly, some algebraical methods of Arbogast.

7.1 The universities and the Royal Society of Edinburgh

Mathematics was taught in all four of the Scottish universities. There were Chairs of Mathematics at Glasgow, Aberdeen, St Andrews and Edinburgh.[1]

On the retirement of Simson in 1761, the teaching at Glasgow was continued by James Williamson (d.1795) who was appointed, as was usual in Scotland, his 'assistant and successor'. The 'assistant and successor' had the duty of teaching and received class fees, while the professor retained the Chair, the house and the stipend. At Glasgow students were not formally required to study mathematics; they could therefore graduate without attending classes in mathematics. This situation did not change until 1826. Consequently, only a small number of students payed the fees to the Professor of Mathematics, and both the Chair and the assistantship were of little economic interest. On 31 December 1765 Williamson attempted to lay down a regulation according to which 'after 10 June of the following year the Professor of Mathematics should examine all candidates for degrees in Arts' (Coutts (1909), p. 308); but his proposal came to nothing. In 1788 Williamson retired from teaching, and in January of the following year James Millar was appointed his assistant and successor. Millar was not a mathematician and preferred to lecture on law. Nonetheless, in 1796 he became Professor of Mathematics and held this post until his death in 1832.

At Marischal College, Aberdeen University, we have already come across Colin Maclaurin (1698–1746) (see chapter 1, section 1.3) and John Stewart (d.1766) (see chapter 4). The latter was Professor of Mathematics for thirty-nine years, from 1727 to 1766. In the second year, Stewart taught arithmetic, geometry, plane trigonometry, practical geometry and elementary algebra, while during the third year he taught spherical trigonometry, conic sections, astronomy and higher algebra. An optional third mathematical class covered advanced algebra, fluxions and parts of Newton's *Principia* (see Gerard (1755)). During Stewart's years of office a reform took place at Marischal: in 1753 regenting was abolished. The curriculum was changed as follows:

According to the accounts from Aberdeen about the middle of December, the Principal, Professors, and Masters of the Marischal College, in order to render the study of the sciences more natural and progressive, and to fit their students to be *useful in life*, have unanimously resolved, that henceforth the following *general order* in teaching shall be observed in that University. 1. That the first year shall be wholly employed in Classic Literature under the Professor of Greek, as formerly, being the indispensable foundation of all scholarship. 2. That the second year of the academic course shall be spent in teaching History, Geography, Chronology, and an introduction to Natural History, commonly called *special physics*; at the same time that the whole students of this class shall attend lessons from the Professor of Mathematics. 3. That the third year be employed in the scientific parts of Natural Philosophy, the Laws of Matter, &c. commonly called *general physics*, – such as Mechanics, Hydrostatics, Pneumatics, Optics, and Astronomy. 4. That the fourth and last year be allowed to the study of abstract sciences, Pneumatology, Morals and Logic, or the Art of Reasoning. And, that henceforth each Professor be employed in cultivating and teaching *one* particular branch of knowledge. (*The Scots Magazine*, XIV (Dec 1752), p. 606)

The teaching of mathematics was continued by William Trail (d.1831), a pupil of Robert Simson, from 1766 to 1776 and by Robert Hamilton (1743–1829) from 1779 to 1824 (for further details see appendix C.7).[2]

Both Trail and Hamilton were competent mathematicians, but it is difficult to assess the level of their teaching. Trail was the author of an elementary treatise on algebra (1796), first published in 1770, and is described by Ponting (1979b), p. 167, as lecturing twelve times a week. R. Hamilton, better known as a political economist, published in (1800) his *Heads of Part of a Course of Mathematics*, in which he stressed the practical applications of mathematics to mensuration, surveying, dialling, navigation and astronomy (Ponting (1979b), p. 168). During the eighteenth century all the professors of mathematics at Marischal taught fluxions only in an optional third class followed by a small number of students.

In eighteenth-century King's College, Aberdeen, the Chair of Mathematics was practically vacant (see appendix C.7).[3] In 1753 King's Senatus opposed the reform which took place in Marischal, therefore during the whole century mathematics was taught by the regents. According to Ponting,

since he taught the same students for three years the regent could and sometimes did allow some mathematics from the tertian course to overflow into the magistrand year. With more time available the student in the Old Town (King's) progressed considerably further in the subject than his New Town (Marischal) counterpart. He studied conic sections in much greater detail and made considerable progress in fluxions (a topic not touched until the optional third class at Marischal), during his tertian year. In 1800 a regent, Robert Eden Scott, Gordon's grandson, prepared extensive notes (MS. K. 178) for his students on Dialling, Conic Sections and Fluxions 'for want of a suitable elementary treatise on these subjects' (Ponting (1979b), p. 172).

At St Andrews the domination of the Gregorys, who had occupied the Chair of Mathematics from 1669, came to an end in 1763. The professorship passed to Nicolas Vilant (d.1807) and then to Robert Haldane (1772–1854). Here mathematics was taught in the first and second years, according to a regulation introduced in 1747. A third, more advanced, class 'is said to have commenced as early as 1793, although no trace of it appears untill 1822' (J. M. Anderson (1905), pp. xxx–xxxi). However, it must be said that mathematics was often taught by private tutors who do not appear in the official records of the universities. In the case of St Andrews this seems to have been particularly relevant. In fact three of the best Scottish mathematicians of this period, John Playfair, John Leslie and James Ivory, were all students at St Andrews. Clearly, in their mathematical studies they went beyond what was required in the curriculum. The last two are said to have studied under the Rev. John West, assistant to the Professor of Mathematics (see *Proceedings of the Royal Society*, IV (1842), p. 406).

At the University of Edinburgh Maclaurin was succeeded in 1747 by

Matthew Stewart, who made important contributions to the study of geometry. His son Dougald Stewart, known today as a philosopher of the common sense school, took over the teaching from 1772 to 1785 when he was appointed to the Chair of Moral Philosophy. In 1785 Adam Ferguson became Professor of Mathematics and John Playfair became joint-Professor (see section 7.2). Playfair moved to the Chair of Natural Philosophy in 1805 when he was substituted by John Leslie in the Chair of Mathematics. Two other names relevant for our historical survey should be mentioned here: William Wallace (see section 7.3), Professor of Mathematics from 1819 to 1838, and John Robison, Professor of Natural Philosophy from 1774 to 1805. This is a remarkable group of mathematicians who, as we will see, were instrumental in improving the teaching of and research into mathematics in Great Britain.

Another important feature of scientific life in Edinburgh was the foundation in 1783 of the Royal Society of Edinburgh. This Society stemmed from a pre-existing Philosophical Society of Edinburgh which was founded in the late 1730s.[4] The Philosophical Society was, in turn, an extension of the Medical Society of Edinburgh. The aim of the founders, who included Maclaurin, was to extend the scope of the Society's interests in order to include natural philosophy. Between 1737 and 1783 the Philosophical Society published only three volumes of *Essays and Observations, Physical and Literary*, in 1754, 1756 and 1771, respectively. In the first volume there is a paper by Stewart on porisms (1754) and two papers by Maclaurin on astronomy, (1754a) and (1754b), while in the second volume Stewart (1756) is concerned with Kepler's problem. The Philosophical Society never reached a level of activity comparable to the Royal Society of London. This is partly explained by the disorders of 1745 and by the death of Maclaurin in 1746, both events occurring only a few years after its foundation. After an encouraging, even if delayed, beginning in the 1750s with the publication of volumes 1 and 2, the Society went into decline until the 1770s. In this decade the bases were laid down for the establishment of the Royal Society of Edinburgh with John Robison as General Secretary. The new Society edited regularly a series of *Transactions of the Royal Society of Edinburgh*, the first volume appearing in 1788. Each volume was divided into a 'Physical' and a 'Literary' class; the former class including 'Mathematics, Natural Philosophy, Chemistry, Medicine, Natural History and whatever relates to the improvements of Arts and Manufactures' (*Transactions of the Royal Society of Edinburgh*, I (1788), p. 12). As we will see in this chapter, many papers, especially those by Playfair, Wallace and Ivory, concerned the calculus of fluxions.

7.2 John Playfair

John Playfair (1748–1819) is today best remembered as a geologist.[5] In fact from 1797 he embarked on the project of systematizing and commenting on James Hutton's theory of the Earth. However, Playfair also occupies an important place in the history of the eighteenth-century British calculus, since he was one of the first to bring continental methods to the attention of British mathematicians.

Playfair entered St Andrews in 1762 with the purpose of qualifying for the Church. He soon displayed a great interest in mathematics, and while still a student lectured on natural philosophy. In 1766, when he was just eighteen years old, he competed for the Chair of Mathematics at Marischal College, Aberdeen: he qualified third after William Trail (d.1831) and Robert Hamilton (1743–1829), who was later appointed to the Chair of Natural Philosophy (1779) and to the Chair of Mathematics (1817) at the same university. In 1772 Playfair was an unsuccessful candidate for the Chair of Natural Philosophy at St Andrews. He had to wait until 1785 for his first appointment, which was as joint-Professor of Mathematics at Edinburgh University. His position as joint-Professor meant that he had to undertake the teaching. Adam Ferguson, the Professor of Mathematics, was in bad health, and in 1785 had exchanged with Dougald Stewart the Chair of Moral Philosophy for that of Mathematics. Playfair's teaching is remembered by Francis Jeffrey as of great importance in up-dating the study of mathematics at Edinburgh:

for the benefit of those who wished to cultivate the higher branches of the science, he taught at intervals a third class, rendered doubly valuable by his intimate and masterly knowledge of the modern analysis, at that time so little attended to in Britain. This class was attended by many who had long finished their academical studies. (Jeffrey (1822), p. xxi)

Unfortunately he did not publish his more advanced lectures on the 'modern analysis'. In (1795) his *Elements of Geometry* were issued, while in (1812–14) a two-volume *Outlines of Natural Philosophy* contained the lectures he gave as Professor of Natural Philosophy.

The commitment of Playfair in extending the boundaries of the fluxional tradition is best displayed in his research papers and his learned reviews of continental works. However, Playfair's first paper (1779) 'On the arithmetic of impossible quantities', which appeared in the *Philosophical Transactions* for the year 1778, showed some signs of attachment to the Scottish geometrical school of Simson. In it Playfair faced the problem of the legitimacy of using complex numbers; it was rejected by some because of the apparent impossibility of interpreting the roots of negatives as geometrical quantities. In fact Playfair began as follows:

The paradoxes which have been introduced into algebra, and remain unknown in geometry, point out a very remarkable difference in the nature of those sciences. The propositions of geometry have never given rise to controversy, nor needed the support of metaphysical discussion. In algebra, on the other hand, the doctrine of negative quantities and its consequences have often perplexed the analyst, and involved him in the most intricate disputations. The cause of this diversity, in sciences which have the same object, must no doubt be sought for in the different modes which they employ to express our ideas. In geometry every magnitude is represented by one of the same kind; lines are represented by a line, and angles by an angle. The genus is always signified by the individual, and a general by one of the particulars which fall under it. By this means all contradiction is avoided, and the geometer is never permitted to reason about the relations of things which do not exist, or cannot be exhibited. In algebra again every magnitude being denoted by an artificial symbol, to which it has no resemblance, is liable, on some occasions, to be neglected, while the symbol may become the sole object of attention. It is not perhaps observed where the connection between them ceases to exist, and the analyst continues to reason about the characters after nothing is left which they can possibly express: if then, in the end, the conclusions which hold only of the characters be transferred to the quantities themselves, obscurity and paradox must of necessity ensue. (Playfair (1779), p. 318)

In this paper Playfair, developing ideas which he probably derived from Lambert's works on trigonometry, maintained that the geometrical analogy between the circle and the hyperbola allowed the use of imaginaries in the expression of circular functions such as

$$\sin(x) = \frac{e^{xi} - e^{-xi}}{2i} \text{ and } \cos(x) = \frac{e^{xi} + e^{-xi}}{2},$$

because of their analogy with the hyperbolic functions

$$\sinh(x) = \frac{e^x - e^{-x}}{2} \text{ and } \cosh(x) = \frac{e^x + e^{-x}}{2}.$$

Naturally these expressions allowed the algebraic proof of many trigonometrical formulas. Playfair continued by showing the utility of complex numbers not only in trigonometry but also in the theory of integration. Playfair's conclusion was, however, old-fashioned, since he restricted the use of complex numbers solely to the expression of circular functions:

1. That the only cases in which imaginary expressions may be put to denote real quantities, are those in which the measures of ratios or of angles are concerned.

2. That the property of either of those measures, so investigated, might have been inferred from analogy alone. (Playfair (1779), p. 335)

Playfair never returned to this thesis.[6] Soon after this paper he became convinced of the necessity to abandon geometrical analogies as justification for the use of mathematical symbols. In doing so, he distanced himself from the Scottish geometrical school of Simson and Stewart, but not from the fluxional tradition which also included, as we have seen in chapter 6, defenders of the 'analytic art' such as Simpson, Landen and Waring. What was new in Playfair was his ability to discern important themes in the recent developments of the continental calculus.

Three years after the publication of the paper on complex numbers, Playfair visited London, where he met the Astronomer Royal, Nevil Maskelyne (1732–1811). His impressions are revealing:

He [Maskelyne] is much attracted to the study of geometry, and I am not sure that he is very deeply versed in the late discoveries of the foreign mathematicians. Indeed, this seems to be somewhat the case with all English mathematicians; they despise their brethren on the Continent, and think that every thing great in science must be for ever confined to the country that has produced Sir Isaac Newton. Dr. Maskelyne, however, is more than almost any of them superior to this prejudice. He is slow in apprehending new truths, but his mind takes a very strong hold of them at last. (Playfair (1822a), I, p. lxxviii)

Despite his determination to approach the methods of the 'brethren on the Continent', Playfair did not carry out any research in mathematics. However, three papers published in the *Transactions of the Royal Society of Edinburgh* need to be mentioned. These works do not aim to present new mathematical results, but rather to apply analytical methods to mechanics. The first work, Playfair (1788b), deals with formulas of corrections of barometrical measurements; the second (1805) with formulas for the measurement of the arc of the meridian; and the third (1812) with the problem of finding, given a certain mass, the shape which attracts more strongly a particle in a given direction. This last paper especially might have afforded a chance to apply the techniques of the calculus of variations; but Playfair missed this opportunity:

On considering the question more nearly, I soon found, though it belongs to a class of problems of considerable difficulty, which the Calculus Variationum is usually employed to resolve, that it nevertheless admits of an easy solution, and one leading to results of remarkable simplicity. (Playfair (1812), p. 187)

As a matter of fact, Playfair never adopted the tools devised by Euler, d'Alembert and Lagrange, which he so much appreciated. However, he was competent enough to understand the importance of the new achievements of the continental calculus.

One of Playfair's most influential papers was a review of Laplace's *Mécanique Céleste*. Playfair (1808) was often praised as one of the first attempts to awaken the interest of British mathematicians in the works of the French school. Playfair not only comments on the contents of the first four volumes of Laplace's work, but also places them in the context of the development of eighteenth-century astronomy and, in the end, adds several considerations to the reasons behind the inferiority of British achievements. The progress of mathematics after Newton and Leibniz is described as depending upon the analytic treatment of trigonometry, the discovery of 'partial differences' or 'partial fluxions', the 'calculus variationum', the integration of new functions and of higher order differential equations. In the field of mechanics Playfair notes the emergence of the principle of virtual velocities and its employment in Lagrange's *Méchanique Analitique* (1788). Then he proceeds to show how Laplace was able to employ these tools in the study of astronomy. Playfair devotes particular attention to the three-body problem and to the theory of tides. A final question is put to the reader:

In the list of the mathematicians and philosophers, to whom that science, for the last sixty or seventy years, has been indebted for its improvements, hardly a name from Great Britain falls to be mentioned. What is the reason of this? (Playfair (1808), pp. 279–80)

Playfair goes on to describe the situation in England and Scotland:

The calculus of the sines was not known in England till within these few years. Of the method of partial differences, no mention, we believe, is yet to be found in any English author, much less the application of it to any investigation. The general methods of integrating differential or fluxionary equations, the criterion of integrability, the properties of homogeneous equations, &c. were all to them unknown; and it could hardly be said, that, in the more difficult parts of the doctrine of Fluxions, any improvement had been made beyond those of the inventor. At the moment when we now write, the treatises of Maclaurin and Simpson, are the best which we have in the fluxionary calculus, though such a vast multitude of improvements have been made by the foreign mathematicians, since the time of their first publication. These are facts, which it is impossible to disguise; and they are of such extent, that a man may be perfectly acquainted with every thing on mathematical learning that has been written in this country, and may yet find himself stopped at the first page of the works of Euler and D'Alembert. He will be stopped, not from the difference of the fluxionary notation, (a difficulty easily overcome), nor from the obscurity of these authors, who are both very clear writers, especially the first of them, but from want of knowing the principles and the methods which they take for granted as known to every mathematical reader.

If we come to works of still greater difficulty, such as the *Méchanique Céleste*, we will venture to say, that the number of those in this island, who can read that work with any tolerable facility, is small indeed. If we reckon two or three in London and the military schools in its vicinity, the same number at each of the two English Universities, and perhaps four in Scotland, we shall not hardly exceed a dozen; and yet we are fully persuaded that our reckoning is beyond the truth. (Playfair (1808), p. 281)

Playfair's review is typical of early nineteenth-century Great Britain.[7] As we will see below, Laplace's *Mécanique Céleste* was the stimulus which led many British mathematicians to study the works of their continental colleagues.

7.3 Wallace and Ivory

Playfair's programme was successfully continued by two of his young protégés, Wallace and Ivory. In this section I will deal with their early work up to 1803. From that date onwards they were to be found working in England, concerned in promoting knowledge of French mathematics.

James Ivory (1765–1842) was the son of a watchmaker.[8] In 1779 he entered St Andrews where he studied under John West, an assistant to the Professor of Mathematics. In 1783 he began a theological course, which he completed in Edinburgh in 1786. He did not pursue his career in the Church but was appointed assistant-teacher in an academy in Dundee. Three years later Ivory became a partner in a flax-spinning company; and in 1804 he was appointed to the Chair of Mathematics at the Royal Military College in Great Marlow. In 1819 he resigned because of bad health and moved to London where he spent the rest of his life.

Ivory's work during his early Scottish period consists of a series of papers published in the *Transactions of the Royal Society of Edinburgh*. Only Ivory (1798) is concerned with the calculus; it deals with the problem of expanding

$$(a^2 + b^2 - 2ab\cos(\theta))^{-\frac{1}{2}}$$

into a series of the form:

$$\sum_{n=0}^{\infty} A_n \cos(n\theta).$$

These researches led him on to consider Legendre polynomials and elliptic integrals. Rather than constituting an attempt to approach continental analysis, this research might have been motivated by Landen (1775).

William Wallace (1768–1843), a self-taught mathematician who gained the esteem of Playfair and Robison for his knowledge of geometry, was able to extend Ivory (1798) in his (1805).[9] Wallace's results had, however, been anticipated by Legendre. As we will see in the next section he later recognized Legendre's priority. The other contributions of Wallace in this period were concerned with porisms. He gained a great reputation as a mathematician: an anonymous (1803b) and a too enthusiastic reviewer compared him with Euler and Lagrange. It was not difficult for him to win an appointment as Mathematics Master at the Royal Military College at Great Marlow. He retired in 1819, the same year as Ivory, and moved to Edinburgh where he occupied the Chair of Mathematics till 1838. As we will see below, during their period at the Military College, Wallace and his colleague Ivory became two of the most influential reformers of the British calculus.[10]

7.4 Glenie and Spence

It is appropriate here to mention two minor Scottish mathematicians whose contributions were considered with interest by their contemporaries.

The first is James Glenie (1750–1817), educated like Playfair, Ivory and Leslie at St Andrews. He then went to the Royal Military Academy at Woolwich where he qualified as a military engineer. He served as an artillery officer in the American War of Independence and taught (c. 1805–10) at the East India Company Royal Military College at Addiscombe. Glenie wrote on gunnery (1776), but also on the binomial theorem and on his 'Antecedental Calculus' (1778), (1789), (1793) and (1798). The aim of Glenie's antecedental calculus was to construct a 'Geometrical Method' which was independent of any 'consideration of Motion and Velocity' and equivalent to the Newtonian calculus of fluxions. He employed the binomial theorem in order to express the incremental quotient of algebraic functions as a power series, then quite simply deleted higher order powers and obtained the equivalent of the Newtonian fluxion, or Leibnizian derivative. Notwithstanding the obvious weakness of Glenie's mathematics, his works were fit to be published in the *Philosophical Transactions* and in the *Transactions of the Royal Society of Edinburgh*: a sign of interest in alternative approaches to the Newtonian fluxional calculus. In fact, algebraical proofs of the binomial theorem became a fashionable exercise in the late eighteenth century. Papers such as Abram Robertson (1795) and (1806), Sewell (1796) or Knight (1816) were all devoted to the algebraical derivation, 'independent from the principles of the fluxional calculus', of the binomial theorem. This interest in algebraization was to bear (as we will see in chapter 9, section 9.5) more substantial fruit with the adoption of the Lagrangian programme.

More interesting is another Scottish outsider, William Spence

(1777–1815).[11] He was born in Greenock where he formed a 'Literary Society' together with his young friends. Even though he showed a remarkable aptitude for study, he was sent to Glasgow to be educated as a manufacturer. Nevertheless, Spence devoted himself to mathematics, and in 1809 he published *An Essay on the Theory of the Various Orders of Logarithmic Transcendents*. This was a very interesting work; in fact, Spence was one of the first British mathematicians to show a marked predilection for the Lagrangian school and particularly for Arbogast (1800). It may be that Spence was influenced by Woodhouse (1803) (see chapter 9, section 9.2). Spence (1809) was intended as a part of a 'Theory of Analytic Functions', an expression which recalls the title of Lagrange (1797). Differential notation was employed throughout the work. A function was defined as 'the analytical expression of the result which certain operations produce on a given quantity, or on any number of quantities' (Spence (1809), p. iii). He put forward ideas very similar to Woodhouse on the independence of 'Analytical Mathematics' from mechanical and geo-metrical interpretations:

It is one thing to learn the science of Analytical Mathematics, and another, to learn its practical applications; and although most of our authors, in this country, have mingled these very different subjects together, it is certainly much to be doubted, whether any advance has thence accrued either to the student or to the science itself. There is surely a material distinction betwixt the art of Numeration and the process of Book-keeping. For the understanding of the latter, the former must be learned; but it is quite unnecessary for the mere student of Arithmetic to make himself master of Merchant's Accounts. In the same manner is the general science of Mathematical Analysis related to Geometry and Mechanics. Both these departments of knowledge are greatly facilitated and enlightened by the modern analysis: but the benefit thus conferred is by no means reciprocal, for the principles and methods of the latter are, of themselves, independent, and may be demonstrated without any foreign aid. In Great Britain, however, we do not seem to have sufficiently weighted the importance of this circumstance. Our analytical treatises consist, in great part, of dissertations relative to Statics, Dynamics, &c.; and before the learner can proceed beyond the threshold of science, his attention is called off to consider the path of a projectile, or the vibrations of a pendulum. It may fairly be asked, what have these subjects to do with analysis? and the only answer that can be given to this question is, that they form some of its numerous applications, and exemplify several of its theories; although no otherwise connected with these theories, than as showing how analytical formulae can be interpreted by the combinations of material bodies, and their reciprocal actions on each other. These, undoubtedly, are important enquiries, but are they not out of place in a work destined to the development of any part of Analytical Mathematics? Analysis is the instrument employed in these researches; and should not the powers of the instrument be the first object of instruction, more especially, when all the difficulties which occur in using it are only to be overcome by a thorough possession of its principles? Our mode of instruction, however, is quite the reverse of this. Our pupils are taught the science by means of its application; and when their minds should be occupied with the contemplation of general methods and operations, they are usually employed on particular processes and results, in which no traces of the operations remain. On the Continent, Analysis is studied as an independent science. Its general principles are first inculcated; and then the pupil is led to the applications; and the effects have been, that while we have remained nearly stationary during the great part of the last century, the most valuable improvements have been added to the science in almost every other part of Europe. The truths of this needs no illustration. Let any person who has studied Mathematics only in British authors, look into the works of the higher Analysts on the Continent, and he will soon perceive that he has still much to learn. (Spence (1809), pp. x–xi)

Ideas very similar to Spence's were to be shared by the Fellows of the Analytical Society in the 1810s (see chapter 9, section 9.5). The logarithmic transcendants with which Spence (1809) was concerned were defined as:

$$L^n(1 \pm x) = \int_0^x \frac{L^{n-1}(1 \pm x)\mathrm{d}x}{x},$$

where $L^1(x) = \ln(x)$. Landen in his (1761) had already studied the properties of these functions. Spence tabulated the functions L^2 and L^3, and applied them to the integration of

$$\int FL^1(V)\mathrm{d}x, \int FL^2(V)\mathrm{d}x,$$

where $F = X/(a + bx \pm x^2)^{\frac{1}{2}}$ and X and V are rational functions. He also had a notation for partial derivatives:

$$^{m,n}\{Z_{x,y}\}^{x,y} = \frac{\partial^{m+n}}{\partial y^m \partial x^n} Z(x,y),$$

and expressed the relations

$$a\frac{\partial^n}{\partial x^n} = \frac{\partial^n}{\partial x^n}a \text{ and } \frac{\partial^2}{\partial x \partial y} = \frac{\partial^2}{\partial y \partial x}.$$

In addition, he treated double integrals, writing (Spence (1809), p. 104):

$$2\int P\mathrm{d}x\mathrm{d}y.$$

Moreover, very simple functional equations were solved by assuming a power series representation of the unknown function.

We may consider Spence (1809) as a bold attempt by an unknown and provincial mathematician to break with the fluxional tradition and to employ the new symbolical techniques of the Lagrangian school. Less fortunate was Spence (1814), in which an attempt was made to provide a global treatment of algebraical and differential equations (another classic topic in the Lagrangian calculus of operators).

Spence died in 1815. His friends sent some of his manuscripts to John Herschel (1792–1871), one of the young promoters of the British Lagrangian school. Herschel edited them and added a preface.[12] William Spence, the President of the Greenock Literary Society, had worked in isolation in a direction which was to dominate the scene of British mathematics in the first half of the nineteenth century.

8

THE MILITARY SCHOOLS (1773–1819)

THE MILITARY schools played an interesting role in the reform of the British calculus. At Woolwich we find a group of engineers including Hutton, Barlow, Gregory and Bonnycastle. They had at their disposal the dockyard, the arsenal and nautical instruments, and so were in a position to investigate the strength of materials, magnetism and ballistics. Their knowledge of similar work in France was extensive, while their own contributions to engineering were known on the continent. They did not excel as mathematicians, and, with the exception of Bonnycastle, they did not break with the fluxional tradition. As teachers they could not introduce any sophisticated innovations into the curriculum for the 'raw and inexperienced' cadets. However, with their textbooks and essays they greatly contributed to improving the knowledge of continental science in Britain. At Sandhurst more basic research on integration was carried out by the two Scots, Ivory and Wallace, assisted by Leybourn and several other colleagues. They contributed to a *Mathematical Repository*, where differential notation, partial derivatives and difference equations appeared as early as 1806. Their work marked a significant step towards the continental calculus. Certainly it was not possible for them to introduce their students to more than algebra and trigonometry. The masters at the military schools also participated actively in the astonishing number of scientific journals and encyclopaedias published during the first decades of the nineteenth century. Their contribution to the development of British science has been unjustly neglected. Some attention is given in this chapter also to the Royal Naval Academy at Portsmouth, founded in 1733 to train naval officers. However, the study of the three military schools treated in this chapter does not at all exhaust the subject of military education in eighteenth-century Britain. Only a small fraction of the officers were trained at Woolwich, Sandhurst or Portsmouth. The great majority were trained directly on the field, or on the quarter-deck: by apprenticeship, so to speak. Furthermore, there was quite a number of private schools and academies which could compete with the three Royal military schools.

8.1 The Royal Military Academy at Woolwich

The Royal Military Academy at Woolwich was instituted in 1741 with the purpose of 'instructing the raw and inexperienced people belonging to the Military branch of this·office, in the several parts of Mathematics necessary to qualify them for the service of the Artillery, and the business of Engineers' (W.D. Jones (1851), p. 1).[1] In the beginning the school was attended not only by young students (called the 'gentlemen cadets'), but also by 'Practitioners Engineers, Officers, Serjeants, Corporals[...] and also all such Bombardiers, Miners, Pontoonmen, Mattrossess, and others of the said Regiment [of Artillery] as have a capacity and inclination to the same' (W.D. Jones (1851), p. 2). Two masters were employed: the Chief Master was John Muller, whom we have already mentioned as a writer on mathematics and fortifications (see chapter 4). Gradually the teaching staff and number of cadets increased, while the habit of teaching officers and the craftsmen of the Regiment of Artillery was abandoned. In 1776 the cadets numbered forty-eight and there were just two masters, while in 1806 the cadets totalled more than 180 and there were almost twenty masters, i.e. a professor of mathematics and seven mathematical assistants, a professor of artillery and fortification with two assistants, two drawing-masters, two French masters, a master for fencing, a master for chemistry, and a master for dancing.

A sizeable group of good mathematicians was employed at Woolwich. The most notable in the period 1741–1838 were John Muller, Thomas Simpson, Charles Hutton, John Bonnycastle, Olynthus Gregory, Peter Barlow and Samuel H. Christie. In fact, Woolwich became a centre for the reform of British science. The practical needs of military engineering demanded sophisticated scientific knowledge: this partly explains why the Woolwich masters were so interested in contemporary continental works. The French military schools in particular became a model for many European countries in the early nineteenth century. As we will see, the mathematics masters at Woolwich did not make any significant contribution to mathematics; nonetheless they stressed the importance of the French achievements and wrote a series of works (dictionaries, encyclopaedias, textbooks) which acquainted the British reader with the methods and results of the continental school.

This does not imply that the teaching at Woolwich reached a very high standard: the little we know about it scarcely encourages such an hypothesis. In the syllabuses for 1764 we find that the Professor of Artillery and Fortification based his teaching on 'the following books, viz. Gregory's Practical Geometry, Vauban's Treatise of Fortifications, Muller's Elements of Fortifications, Muller's Attack and Defence of Fortified Places' (Manners (1764), p. 7).[2] The course on mathematics included algebra, geometry, mensuration of superficies and solids, plane trigonometry, conic sections, theory of perspective, geography and the use of globes. The set texts were 'Selects parts of Professor Saunderson's Elements of Algebra, including common arithmetic, Simpson's Elements of Geometry, Hawney's Mensuration of superficies and solids, Kirby's Theory of Perspective, Cowley's Theory of Perspective demonstrated, Salmon's Geography, Harris on the use of Globes' (Manners (1764), p. 8).[3] The syllabuses for 1776 only differ to a slight extent. In 1792, however, we find a more structured course of mathematics, the result of almost twenty years of Hutton's teaching. The subjects to be studied following Hutton's manuscript *A Course of Mathematics*, 2 vols (1798, 1801) were: arithmetic, logarithms, geometry, algebra, trigonometry, mensuration, conic sections, mechanics, fluxions, hydrostatics and hydraulics, pneumatics, resistance of fluids and gunnery (see W.D. Jones (1851), p. 88). Hutton (1798, 1801) ran to twelve editions, several American editions and even a translation into Arabic. Hutton's *Course* was abandoned in Woolwich only in the mid-1830s, when Samuel Christie introduced the continental differential and integral calculus.

From 1792 to 1806 six new mathematics masters were appointed. Mathematics clearly dominated the course of studies for cadets, but, as can be inferred from the syllabuses, only the most elementary aspects of mathematics were included in the curriculum. Furthermore, we suspect that even the very elementary level required was not reached: from the *Records of the Royal Military Academy*, W.D. Jones (1851), one gets the strong impression that the discipline of both the masters and the cadets was not exemplary. The cadets, who entered at the age of about fourteen, were supposed to leave Woolwich after two years. In fact, after a two-year course a cadet was reported 'fit for a public examination for a Commission in the Royal Corps of Artillery and Engineers' (W.D. Jones (1851), p. 33). Smyth says that 'the first public recorded examination took place in 1765 in the presence of the Marquis of Granby, Master-General' and that public passing-out examinations were instituted in place of 'somewhat perfunctory' viva-voce examinations (see Smyth (1961), pp. 37–8).

8.2 Charles Hutton

Charles Hutton (1737–1823) was born in Newcastle in a family of colliery workers.[4] After being educated at local schools, he was able to set up his own course in the late 1750s. Hutton's flair for teaching was soon revealed: he established himself as one of the most successful mathematics teachers in the region. In 1770 he was asked by the Mayor and Corporation of Newcastle to prepare a survey of Newcastle. When in 1771 the bridge over the Tyne collapsed, Hutton wrote a book (1772) on the stability of bridges. In 1773 the main event in Hutton's career occurred. The Chair of Mathematics at Woolwich became vacant, and a public examination was held for the election of the new Professor: amongst the examiners there were Samuel Horsley, John Landen and Nevil Maskelyne. Hutton was appointed in May 1773 and retained the job until 1807. In this period Hutton produced a massive amount of work. Hutton (1775b) was a five-volume edition of the *Ladies' Diary* from the first issue in 1704 to the 1773 issue. Hutton was the editor of this famous periodical from 1773/4 to 1817 (see section 8.5). In addition, he prepared an abridgement of the *Philosophical Transactions* from 1665 to 1800, wrote his *Mathematical and Philosophical Dictionary* (1796, 1795), two volumes of papers on series and gunnery (1786) and a two-volume *Course of Mathematics* (1798, 1801), and also found time to publish numerical tables.

In 1779 Hutton was elected Foreign Secretary of the Royal Society; it is not surprising that his publishing career and his teaching at Woolwich did not allow him much time to perform the duties of this additional post. In 1783 he was obliged to resign, after a committee had reported on his case. A squabble ensued between the 'mathematicians', Horsley and Maskelyne, and the defender of the 'disciples of Linneus' and President of the Royal Society, Joseph Banks. Banks succeeded in imposing his will, and Hutton did not publish in the *Philosophical Transactions* until after Banks's death. It is revealing that his (1790) appeared in the *Transactions of the Royal Society of Edinburgh*.[5] Hutton retired from teaching in 1807. He spent the rest of his life in Woolwich living on a state pension.

Hutton's researches centred on the convergence of series, and experiments in ballistics, the building of bridges and measuring the Earth's density. His research into series shows a certain influence of Euler in the approach to the problem of convergence (see chapter 9, section 9.5). The formal use of divergent series was accepted by Hutton as a legitimate method. Hutton referred very often to Euler with great esteem; the article 'Euler' in Hutton's *Dictionary* (1796, 1795) is very detailed and appreciative.

Hutton did not hide his admiration for analytical methods and his belief
that geometrical constraints are extraneous to research. He concluded the
article 'Analysis' as follows:

if we should look no farther than the method of the ancients, it is probable that,
even with the best genius, we should have made but few or small discoveries, in
comparison with those obtained by means of the modern analysis.[...] Upon the
whole therefore, the state of the comparison seems to be this; That the method of
the ancients is fittest to begin our studies with, to form the mind and to establish
proper habits; and that of the moderns to succeed, by extending our views beyond
the present limits, and enabling us to make new discoveries and improvements.
(Hutton (1796, 1795), I, p. 107)

Hutton's *Dictionary* reveals a deep and extensive knowledge of continental
works and is still used by historians as a valuable source. The British
reader was provided with bibliographical and biographical information
which could be used to orientate him in the variegated world of
continental mathematics. Hutton gave space to d'Alembert, Euler and
Lagrange, providing an outline of their methods and their results.
However, Hutton's *Dictionary* did not include technical details of the
foreign works. It is revealing that the brief entries for 'differential' and
'integral' referred to 'fluxions' and 'fluents'. The explanation for this last
entry went as far as the integration of rational functions, but almost
nothing was said about differential equations. The chapter concerned with
fluxions of Hutton's *Course* (1798, 1801) was also fully within the
fluxional tradition. Hutton's achievement is in a way analogous to that of
Playfair. Both advocated the 'new analysis' and gave publicity to foreign
mathematics, but were unable to use it in research and, in reality, never
even attempted to teach it in written works.

8.3 Olynthus Gregory and Peter Barlow

The *Ladies' Diary* was a vehicle for promising young mathematicians to
make a name for themselves among the wide and diverse circle of
philomaths. Charles Hutton, as editor of the *Ladies' Diary*, was particularly
impressed by the answers of two contributors: Peter Barlow (1776–1862)
and Olynthus Gilbert Gregory (1774–1841). We know very little of their
early lives and education. It seems that they both ran schools and that
Gregory spent some time as a bookseller in Cambridge. When the staff of
the Royal Military Academy was increased, they were appointed
'mathematical assistants', most probably through Hutton's influence.
Gregory was elected in 1803 and became Professor in 1821, while Barlow
was elected in 1806.

Gregory prepared a course which consisted of a treatise on astronomy (1802), a treatise on mechanics (1806), and a translation (Haüy (1807)) of Haüy's *Traité Elémentaire de Physique*. The mathematical basis to Gregory's course is very weak, but it is still interesting how he updated the teaching of 'natural philosophy' by referring to French works. In his *Mechanics* attention is paid to Lagrange (1788) and Carnot's *Géométrie de Position* (1803). When treating floating bodies, he refers to Euler's *Scientia Navalis* in the translation (1776) of Henry Watson. In book III on hydrostatics, hydrodynamics and pneumatics he employs de Prony's *Architecture Hydraulique* (1790–6). A faithful pupil of Hutton, he continued his master's efforts at spreading knowledge of French physics in Great Britain. Surprisingly, his attitude towards the fluxional calculus was more conservative than Hutton's; he wrote:

The Editor has long been of the opinion that, in point of intellectual conviction and certainty, the fluxional calculus is decidedly superior to the differential and integral calculus. (Hutton, *Course of Mathematics*, 11th edn. (ed. Olynthus Gregory), 2 vols (1836–7), II, p. 203. Quoted in Howson (1982), p. 251)

After referring to D.M. Peacock, one of the opponents of the Analytical Society of Cambridge (see chapter 9, section 9.5), he added that to think of calculus 'without motion' was akin to thinking of 'war without bloodshed, gardening without spades' (see Howson (1982), p. 251).

Peter Barlow's main contributions are his numerical tables (1814b) and his researches into the strength of materials (1817), optics, and magnetism (1820). These last gained him an international reputation. He studied the compass deviation caused by the iron in ships and devised a method of correcting it by placing a small iron plate close to the compass. Poisson in 1824 supplied a mathematical theory to explain Barlow's experimental results. Barlow contributed to the Rees' *Cyclopaedia* (1802–20) with articles on algebra, analysis, geometry, strength of materials and to the *Encyclopaedia Metropolitana* (Division II, Volume I (1829)) with lengthy essays on mechanics, hydrodynamics, astronomy and magnetism. In these articles he showed all his profound acquaintance with foreign works, but he still used the fluxional notation. In line with the research carried out by Hutton into ballistics and bridge construction, Barlow may be considered more an engineer than a mathematician. And it was as an engineer that he was able to contribute to the researches of the French school which he desired to make known to his fellow countrymen. The introduction of French mathematics into Great Britain, that we are going to describe below in this chapter and in chapter 9, can be seen as part of a more general interest in French science, which was also stimulated by the Woolwich mathematics masters.

8.4 The Royal Military College at Sandhurst

The Royal Military College was instituted in 1799.[6] It was divided into two 'departments'. A Senior Department for the instruction of a small number (about forty) of army officers was formed in 1799/1800 and was based at High Wycombe, in Buckinghamshire. The Junior Department, formed in 1802, was attended by young students and was based near to Great Marlow. The two departments were moved in 1813 to Farnham and Sandhurst in Surrey, respectively. The Junior Department was much larger than the Senior Department: to begin with there were just sixteen cadets, but a Royal Warrant of 17 April 1803 fixed the number at 400. 412 was the number indicated in the Royal Warrant of 27 May 1809. This level was never actually reached: in 1804 there were 199 cadets, while in 1809 the cadets totalled 320, the highest recorded number I have seen. After this year it seems that the students decreased (e.g. in 1824 there were 212).

The level of instruction given to the cadets during the first twenty years of the College's activity can be inferred from indirect sources and from Dalby's *Course of Mathematics* (1806) 'designed for the use of the officers and cadets in the Royal Military College'.[7] Dalby (1806) treats, in a simplified way, arithmetic, geometry, mensuration, algebra, conic sections, mechanics (the six simple machines), hydrostatics, hydraulics and pneumatics. It was definitely more elementary than Hutton (1798, 1801), the analogous text used at Woolwich. We know that around 1820 a cadet had to possess the following qualifications before leaving Sandhurst:

Thorough knowledge of Euclid, Books 1–6; well versed in either Classics, French, German, or History, conversant with the 1st and 3rd systems of Vauban (on fortifications); proficient in Military Drawing; general conduct impeccable. (Mockler-Ferryman (1900), p. 23)

What is known about the Senior Department? Is it true that the few officers attending it were required to follow an advanced course in, for example, ballistics, embankment construction and cartography? The syllabus for 1802 does not support this view, but that of 1849, according to an 'experienced officer', included the differential and the integral calculus, dynamics and statics, and 'practical astronomy' (see anonymous (1849), p. 55). Questions about the teaching of mathematics at Sandhurst are difficult to answer but not inappropriate, since the Military College had on its staff two of the best British mathematicians of the period. Wallace was appointed mathematics master in 1803, while Ivory was taken on in 1804. They both resigned in 1819. To them we should add Thomas

Leybourn (1770–1840), a mathematics writer and editor of mathematical periodicals (see below), who was at Sandhurst from 1802 to 1839.

8.5 The *Ladies' Diary* and the *Mathematical Repository*

The Military Schools of Woolwich and Sandhurst are connected with the publication of two mathematical serials: respectively, the *Ladies' Diary* and the *Mathematical Repository*.[8]

The former was launched in 1704 and continued up to 1840. It was the most famous mathematical periodical of the eighteenth century 'philomaths'. Many fluxionists, such as Simpson, Landen and Hutton, began their careers as contributors to the *Ladies' Diary*.

The first issues were concerned with enigmas, bad poetry, puns and other miscellanies; however, from 1707 the *Ladies' Diary* included a section on mathematical questions to be answered in the following issue. At the beginning the questions were very easy, but little by little they increased in number and difficulty. By 1840 more than 1800 questions had appeared. The connection with Woolwich began with Thomas Simpson who edited the *Diary* from 1754 to 1760. After thirteen years the *Diary* came back to Woolwich with Charles Hutton, editor from 1773/4 to 1817, and with Olynthus Gregory, editor from 1818 to 1840.

The *Diary* reflects very faithfully the level of instruction and mathematical expertise of the British philomaths, and it is a good guide for estimating their number. No less than thirty imitations were launched before 1800, an indication of the success of such publications. It seems that the *Ladies' Diary* was able to survive this competition because its editors, especially after Simpson, were careful to maintain respectable scientific standards. In the 1740s the mathematical questions became more difficult and the answers less cavalier in near coincidence with the Berkeley dispute, Maclaurin's *Treatise* and the increase in textbooks on fluxions which has been noted in chapter 4. As we have already seen, the calculus of fluxions had a very restricted circulation in the first decades of the eighteenth century; not surprisingly, answers involving the use of fluxions were very rare before 1740, and there were in fact only three before 1730 (all on finding maxima or minima).[9] Later the mathematics of the *Ladies' Diary* included integration and series. It is interesting to see that the number of contributors who were able to tackle these more advanced topics was quite limited (from 1740 to 1773 their number can be estimated at thirty). Moreover, most eighteenth-century philomaths did not touch the integration of $\sin(x)$; neither can any great results or deep discussions be found in the answers of the thirty top correspondents of the

Ladies' Diary. Good mathematicians such as Landen and Simpson contributed, but if they had something interesting to put forward they turned to the *Philosophical Transactions* or to an independent publication.

The *Ladies' Diary* was important as a vehicle of information for the philomaths, but was not a research journal. J. Orchard, 'Writing-Master and Teacher of the Mathematics at Gasport' (Hutton (1775b), II, p. 333), found it useful to publish answers in order to advertise his school. A young self-taught mathematician might find an audience to appreciate his ability. This kind of contributor always signed his name clearly and in full, indicating his place of activity and profession. Important figures preferred to conceal their identity behind pen-names. For example, Simpson, who when young and unknown signed his name and surname, later preferred to attach to his answers pseudonyms such as 'Hurlothrumbo, Kubernetes, Patrick O'Cavannah, Anthony Shallow, Timothy Doodle, Marmaduke Hodgson' (Leybourn (1817), I, p. ix).

The Sandhurst equivalent of the *Ladies' Diary* was quite different. It was launched by Thomas Leybourn in 1795 with the title *The Mathematical and Philosophical Repository*. After 1804 it continued in a new series divided into six volumes. This publication, especially the first three volumes of the new series (1806, 1809, 1814), is one of the most important works in the reform of the British calculus. Each volume included a section of questions and answers, a section of original essays and a section of 'mathematical memoirs extracted from works of eminence'. Spread throughout the volumes one also can find brief reports on British and foreign (especially French) publications, and, from volume three, a publication of the Senate House Examinations of Cambridge University.

In comparison with the *Ladies' Diary*, Leybourn's *Repository* was the work of a small number of people. The authors were mainly Leybourn himself, William Wallace, James Ivory and their colleagues at the Royal Military College: John Lowry, James Cunliffe and Mark Noble. Answers and original essays also arrived from Woolwich, written by Barlow and Gregory. A group of outsiders made contributions: they provided at a rough estimate about one-quarter of the answers and very few essays. Among them the most notable were Thomas Knight (binomial theorem and series), John Toplis (1774–1857) and Benjamin Gompertz (1779–1865). The best work, however, was done by Ivory and Wallace, and the Sandhurst men wrote a large part of the *Repository*. No matter what the cadets studied, mathematics was cultivated with enthusiasm at the Royal Military College. The level of the questions and especially of the essays was very high and marked a point of departure from the fluxional tradition. Notably, the differential notation was employed by Ivory and

Wallace as early as 1807/8 (see Leybourn (1806–35), II (1809), pp. 67–72, 118–24).

Another work rendered in differential notation was Wallace's translation of Legendre's 'Mémoire sur les transcendantes élliptiques' (1794).[10] In the years immediately preceding Wallace's translation (which appeared in vol. II (1809) and vol. III (1814)), the rectification of the ellipse had received attention from Landen (1775), Ivory (1798), Hellins (1798b), (1800) and (1802), Woodhouse (1804) and Wallace himself (1805). The translation of Legendre (1794), in which the elliptic integrals were classified, as nowadays, into three species, was motivated by the great interest of the British mathematicians in this aspect of the theory of integration.[11] All the results of the British could be derived from Legendre's far more general treatment. Questions on elliptic integrals had already occupied Ivory and Wallace in the first volume of the *Mathematical Repository* (1806), where reference was made to Legendre (1794), which therefore became known to the Sandhurst men in between 1804 and 1806 (see Leybourn (1806–35), I part 1 (1806), pp. 34–42, 153–4). The fact that Ivory and Wallace began using the differential notation after this date suggests that their total conversion to the continental calculus, inspired by Playfair, was reinforced by their encounter with a work which solved and systematized all their problems on elliptic integrals.

Other points of interest in the *Repository* are Ivory's use of 'partial fluxions' and Wallace's integration of finite difference equations (see Leybourn (1806–35), II part 3 (1809), pp. 122–4, 156–9). At first, the other contributors did not follow Ivory and Wallace in using the continental notation; nonetheless many of their solutions and essays deserve attention since they often went deeper into a subject than the philomaths of the *Ladies' Diary*.

8.6 Ivory's break-through

Laplace's *Mécanique Céleste* provided a stimulus stronger than Legendre (1794) to abandon the fluxional calculus. The impact of this work on British science cannot be underestimated. It was immediately recognized by many British mathematicians as the masterpiece which crowned Newtonian mechanics and astronomy. In particular, planetary motions and the Earth's shape were regarded as the most important objects of science in Great Britain as well as on the continent. The achievements of Laplace in these fields were outstanding: an urgent need arose to understand his work.

Playfair advertised Laplace's *Mécanique* in (1808), while in 1814 there appeared John Toplis's translation of the first two books. Playfair's lectures as Professor of Natural Philosophy (1812–14) constituted an easy introduction to the results obtained by Laplace on the stability of the solar system and the tides (Playfair (1812–14), II, pp. 229–341). These were some of the early efforts made to promote the knowledge of Laplace's masterpiece.

Ivory was one of the first British mathematicians to follow up Playfair's suggestions. In (1809) Ivory concerned himself with Laplace's treatment in Book 3 of the *Mécanique* of the attraction of homogeneous spheroids. In this paper Ivory demonstrated his theorem on the attraction of confocal ellipsoids. He stated it as follows:

If two ellipsoids of the same homogeneous matter have the same excentricities, and their principal sections in the same planes; the attractions which one of the ellipsoids exerts upon a point in the surface of the other, perpendicularly to the planes of the principal sections, will be to the attractions which the second ellipsoid exerts upon the corresponding point in the surface of the first, perpendicularly to the same planes, in the direct proportions of the surfaces, or areas, of the principal sections to which the attractions are perpendicular. (Ivory (1809), p. 355)

This theorem can be derived from two of Laplace's results. The first says that the potentials of confocal ellipsoids at an exterior point are proportional to their masses. The latter says that the attractive force of an ellipsoid at a point on an axis is a linear function of the length of that axis. Ivory's theorem is an important one, for it played a role in the development of potential theory. It was noted by Legendre, Poisson and Gauss. This meant that Ivory had attained a result which received the attention of the best continental mathematicians. Furthermore, Ivory employed in his paper (1809) the differential notation, Euler's bracketed notation for partial derivatives, and repeated integral signs for multiple integrals. Legendre's functions were used throughout the paper and functions were expanded into infinite series without great concern for convergence. His understanding of the calculus was clearly differentialist: e.g. he used $dxdydz$ to represent an infinitesimal cube. Ivory continued this trend of research in a sequel of works on attraction, fluid equilibrium and physical astronomy. He can be regarded as one of the first British mathematicians to free himself fully from the fluxional tradition.

8.7 Other journals and the encyclopaedias

Two other subjects must be considered briefly at this point; i.e. the launching of new scientific journals and the publication of encyclopaedias. At the turn of the century British science was characterized by an impressive series of encyclopaedic publications (such as Hutton's *Dictionary*) and scientific periodicals (such as Leybourn's *Repository*) which included not only material for popular consumption, but also high-level research. One of the aims of this kind of literature was to introduce French science into Great Britain: attention was paid to the amateur as well as to the researcher. Original papers written by British authors were included, but many papers were just translations of papers appearing in French journals.

The main periodicals launched with this purpose were the *Philosophical Magazine* (1798–1826) edited by Alexander Tilloch (1823–4 by Tilloch and R. Taylor, 1825–6 by R. Taylor), the *Journal of Natural Philosophy, Chemistry and the Arts* (1797–1802, n.s. 1802–13) edited by William Nicholson and the *Annals of Philosophy* (1813–20) edited by Thomas Thomson. The Royal Institution was responsible for the launching of the short-lived *Journal of the Royal Institution* (1802–3) and the *Journal of Science and the Arts* (1816–19). I will treat these periodicals together because they contained little mathematics, being largely devoted to chemistry. In fact Thomas Thomson describes the contents of his *Annals* as follows:

It has been complained that too great a proportion of the *Annals* has been dedicated to Chemistry. We admit that, like all other journals of the present day, our *Annals* must contain a greater proportion of chemistry, which is making a rapid progress, than of those sciences which are in a great measure stationary. But any person who will run over the contents of our volume, will find essays belonging to the following branches of knowledge, namely, Agriculture, Anatomy, Astronomy, Biography, Botany, Geognosi, Hydraulics, Magnetism, Medicine, Meteorology, Mineralogy, Optics, Physiology, Statistics. (*Annals of Philosophy*, I (January–June 1813), p. iv)

Taking into account that the astronomical and the statistical papers were generally about positional astronomy and tables of mortality, one might conclude that mathematics was altogether absent from the *Annals* and the other similar periodicals. But this is not the case. For instance, the *Annals* published a translation of Delambre's reports on the mathematical department of the French *Institut*. A translation of Carnot's *Réflections* appeared in the *Philosophical Magazine* in 1800–1; the differential notation was not changed into the Newtonian dots, even though the translator adopted a position in favour of the Newtonian notation.[12] Furthermore, all these periodicals included biographies and obituaries of foreign mathematicians as well as up-to-date information about mathematical works published on the continent. A study of the introduction of French mathematics into Great Britain in the late eighteenth century and early nineteenth century should take into consideration the contribution of the editors of these periodicals in spreading information on French science. However, as we have seen, it is the Sandhurst-based *Mathematical Repository* which played the leading role in the reform of the calculus.

We now turn our attention to another important source of popularization: the encyclopaedias of the first half of the nineteenth century. Among these works, the following are particularly worthy of consideration: the first five editions of the *Britannica* (1768–71, 1778–83, 1797–1801, 1810, 1817), the *Cyclopaedia* edited by A. Rees between 1802 and 1820, and the *Pantologia* edited by J. M. Good, N. Bosworth and O. Gregory between 1808 and 1813. The first parts of the *Edinburgh Encyclopaedia* edited by David Brewster began to appear in 1807/8 and it was completed in 1830. Some of the articles contained in these works are, in reality, long and elaborate essays. For our purposes it is important to note a few of them.[13]

Firstly, there is Wallace's article 'Fluxions' (1810) in the fourth edition of the *Britannica*.[14] This was one of the best treatises on the calculus published in Great Britain before the translation of Lacroix's *Traité Elémentaire* in (1816). Wallace introduced the calculus as a theory dealing with functions developable into Taylor series. But he mixed this Lagrangian approach with a limit definition of the fluxion. He ended his article with a fairly systematic treatment of fluxional equations, which cannot be found in any preceding British treatise. He gave general methods of integration for linear first order fluxional equations, and for some classes of higher order fluxional equations. He included also the criterion of integrability of 'complete' fluxional equations.

Later Wallace published a long article (1815) entitled 'Fluxions' in the *Edinburgh Encyclopaedia*.[15] This was similar in content to his article in the *Britannica*, but it was written in differential notation and included a long bibliography of foreign and British works on the calculus. Wallace (1815) is in fact the first complete English treatise on the calculus written in differential notation. In introducing the differential notation into a treatise on the calculus, this long article, written by the Professor of Mathematics at the Royal Military College at Sandhurst, anticipated by one year the translation of Lacroix's *Traité Elémentaire* (see chapter 9, section 9.5) undertaken at Cambridge by Babbage, Herschel and Peacock.[16]

Another important article, 'Function', in differential notation appeared in 1810 in the Rees's *Cyclopaedia*. Probably written by John Bonnycastle, Master of Mathematics at the Royal Military Academy in Woolwich, it dealt with the Lagrangian foundation of the calculus, already adopted by

Spence (chapter 7, section 7.5) and Woodhouse (chapter 9, section 9.2).[17] Woodhouse was described as a 'learned analyst'. The notation adopted was that used in Arbogast (1800). This article also included a treatment of the calculus of variations in Lagrangian form. Bonnycastle's receptiveness towards the Lagrangian approach to the calculus is also evident from his *Algebra* (1813). For instance, when treating the binomial theorem, Bonnycastle cited Woodhouse (1803) and affirmed that a proof of this theorem should employ only the rules of algebra rather than the method of increments, the calculus of fluxions or some other 'high origin' (Bonnycastle (1813), 2nd edn., II, p. 169). Furthermore, in the chapter 'Functions' he gave an algebraical 'demonstration' of Taylor's theorem explicitly based on Lagrange (1797) (Bonnycastle (1813), 2nd edn., II, pp. 308–21).

Other important articles on engineering, mechanics and astronomy were written by the professors at the military schools (e.g. Barlow's long articles in the *Metropolitana*, Gregory's in the *Pantologia*) and by the Scots (especially Robison and Playfair) in the early nineteenth-century encyclopaedias. Sandhurst, Woolwich and Edinburgh were the main centres for the writing of these essays, and this confirms the importance of these three 'schools' in the reform of pure and applied mathematics in Great Britain.

8.8 The Royal Naval Academy at Portsmouth

We cannot end this chapter without adding some information on the naval counterpart to Woolwich. The Royal Naval Academy at Portsmouth was founded in 1733 to instruct forty 'sons of noblemen and gentlemen' who entered at the age of thirteen. The Academy was attended by very few students; for instance, in 1773 there were only fifteen. It seems that in the Navy a system of patronage prevailed which gave to the Captains the right to have a 'retinue' on board from which the future officers were recruited. This system gave to the Captains several privileges; not least that of cashing the money intended for the Volunteers forming the 'retinue'. In 1806 a reform was attempted. The Royal Naval Academy became the Royal Naval College: the number of students was increased. In 1816 it was established that there should have been one hundred during war-time and eighty in times of peace. But this attempt at reform failed. In 1829 the school was opened to train some commissioned officers, and ten years later it became an academy for adult education.

The system of education at Portsmouth was indeed peculiar. In the two years of their permanence in the Academy the students had to copy in good handwriting a 'Plan of Learning' which included navigation, geometry, arithmetic, English, French, drawing, fencing and dancing: once this was finished and illuminated the boy could quit his *alma mater* holding the following certificate:

Mr.————has, in ————years————months, and ————days, finished the Plan of Mathematical Learning, and made a manuscript copy thereof: in consequence, he is judged qualified to serve H.M.Navy. (Quoted in Lewis (1939), p. 89)

When the 1806 reform occurred a new Head-Master was appointed: James Inman (1776–1859). Educated at Sedbergh Grammar School as a pupil of John Dawson (1734–1820) (see chapter 5, note 1), he entered St John's College, Cambridge, in 1794 and graduated B.A. in 1800. Subsequently he worked as astronomer on board of HMS *Investigator*. He wrote several books on navigation. Particularly fortunate were his *Nautical Tables...*, (London, 1829) and his translation of Frederik Henrik Chapman's *Treatise on Ship-Building...*, (Cambridge, 1820). In fact Inman established in 1810 a school of naval architecture. He also wrote a textbook to be used in the College entitled *The System of Mathematical Education...*, (Portsea, 2 vols, 1810, 1812). Even though this work is very elementary (it covers only elementary algebra and plane geometry), it testifies that Inman was trying to change the previous system based on the 'Plan of Studies'. In fact he prepared the following curriculum for two senior classes:

Fifth half-year: Fortifications, doctrines of projectiles and its application to gunnery: principles of fluxions and applications to the measurements of surfaces and solids: generation of various curves, resistance of moving bodies, mechanics, hydrostatics, optics, naval history and nautical discoveries.
Sixth half-year: More difficult problems in astronomy, motions of heavenly bodies, tides, lunar irregularities: the 'Principia' and other parts of Newton's Philosophy to those sufficiently advanced. (Lewis (1939), p. 91–2)

This was too ambitious a target, taking into consideration the starting point. Michael Lewis comments:

How did they do it? The age of the eldest scholar, we must remember, was somewhere between fourteen and a half and fifteen. So it would really be more apposite to ask, *Did they do it?* (Lewis (1939), p. 92)

Two other mathematics masters at Portsmouth should be mentioned here. John Robertson (1712–76), appointed in 1748 Master of the Mathematical School at Christ's Hospital and in 1755 Master of the Royal Naval Academy at Portsmouth. In 1766 he retired to become clerk and librarian of the Royal Society. He was the author of *The Elements of Navigation*... (1754). George Witchell (d.1785) was appointed Master at Portsmouth in 1767. He is remembered in Cotter (1968), p. 226, as the inventor of a method for clearing lunar distances published in the *Nautical Almanac* for 1772.[18]

9

CAMBRIDGE AND DUBLIN (1790–1820)

IN THIS final chapter we will consider the important contributions to the reform of the calculus which took place in Cambridge and Dublin. There is a link between the two universities, since John Brinkley, one of the most influential Dublin reformers, was educated at Cambridge and brought the heritage of Maskelyne and Waring to Ireland. Furthermore, both the Dublin and the Cambridge reformers were deeply concerned with the teaching of mathematics. Their attempt to reform the teaching of mathematics (and the calculus in particular) was much bolder than that of Playfair in Edinburgh, while – as we have seen – in the military schools such a project could not be implemented. The Dublin group insisted more on the teaching of applied mathematics (mechanics, physical astronomy, optics, etc.), whereas the Cambridge group was definitely purist–algebraist. A distinction must also be made at the level of research: in Ireland the stimulus came from Laplace, while in Cambridge the reformers were followers of Lagrange. Scholars of William Rowan Hamilton's optics, quaternions and mechanics, as well as scholars of the algebras of Peacock, Boole and De Morgan, will find this distinction quite significant.

9.1 Fluxions in Cambridge

During the late eighteenth century, Cambridge did not appear a promising centre of mathematical reform. Notwithstanding the fact that mathematics had become the most important subject in the education and in the ranking of students, their curriculum did not include any of the advances made after the 1720s. Research was not encouraged: mathematics was seen merely as a selective discipline which helped to develop the students' powers of reasoning.[1]

Samuel Vince (1749–1821) and William Dealtry (1775–1847) are representative of this conservative trend. Vince, the son of a bricklayer, was given local support to study at Cambridge. He graduated B.A. as a Smith's prizeman in 1775 and obtained an M.A. three years later. His main contributions were in mechanics and astronomy: as a mathematician he had a passing interest in the summation of series (1783), (1785) and integration (1786). In 1796 he succeeded Anthony Shepherd as Plumian Professor of Astronomy.[2]

With his colleague James Wood (1760–1839), who later became master of St John's, Vince wrote *The Principles of Mathematics and Natural Philosophy* (Vince and Wood (1795–9)) in four volumes: Wood wrote the sections on algebra, mechanics (1796) and optics. Vince wrote the sections on fluxions (1795), hydrostatics and astronomy. As was usual, mathematics was not employed in the treatment of mechanics. The mechanics was somewhat in the style of Parkinson (1785), a textbook already in use in Cambridge, in which the three Newtonian laws were employed in a descriptive study of the 'six machines', pendulums and Newton's three laws of motion. The optics reproduced some results of Smith (1738): the treatment was confined almost exclusively to geometrical optics and a description of optical instruments. The hydrostatics was in part based on the early eighteenth-century Saunderson's lectures,[3] but Hamilton (1774) and Parkinson (1785) were also considered, as well as Jurin's researches into capillary tubes, (1744a) and (1744b), and Atwood's experimental lectures (1776). The astronomy dealt with positional and observational aspects and enabled the student to record the position of planets and stars in the sky. But no attempt was made to introduce a mathematical treatment of physical astronomy.

Vince is also the author of a much larger astronomy text in three bulky volumes (1797, 1799, 1808) in which he showed a very good knowledge of contemporary work in the field. This text was not a work for students but for professional astronomers, and it proved to be useful for the working astronomer. From the point of view of mathematics, however, it did not surpass the smaller treatise on astronomy published in Vince and Wood (1795–9). Vince concerned himself with astronomical tables but not with the underlying physics and mathematics.

Vince's *Principles of Fluxions* (1795) found a natural place within the scarcely original course 'designed for the use of students in the University'. A comparison with mid-century treatises would show that there was nothing new to be found in Vince's textbook. But the system of education and examination in Cambridge did not require any novelty. In fact, treatises such as Maclaurin (1742) or Simpson (1750c) are much more interesting since they include original research, while Emerson (1743) could be read as an advanced continuation of Vince's *Principles of Fluxions*

(1795). It seems clear to me that the potential readers Vince and Wood had in mind were those among the Cambridge students who were ambitious enough to desire a first class in the Tripos.

William Dealtry (1775–1847), a Cambridge graduate who was several times moderator at the examinations and taught mathematics from 1808 to 1813 at the East India Company College, Haileybury, undertook the task of making a copy. His textbook on fluxions, Dealtry (1810), not only had the same title as Vince (1795) but the contents were identical. A comparison of the two would show that many 'Examples' are taken word for word, with the same x's, z's and y's, from Vince (1795). Dealtry (1810) was the last treatise on fluxions. It ran to a second edition in 1816 while the fifth edition of Vince (1795) appeared in 1818.[4]

9.2 Robert Woodhouse

Edward Waring, the Lucasian Professor, had died in 1798 a lonely and misunderstood man who exerted very little influence on research into the calculus and who was practically ineffective as a teacher. After him the next notable mathematician in Cambridge was Robert Woodhouse (1773–1827), the son of a linen draper, who graduated B.A. in 1795 and M.A. in 1798. He remained at Cambridge as a fellow of Gonville and Caius College and was for several years Moderator of the Tripos. He was appointed Lucasian Professor in 1820 and Plumian Professor of Astronomy and Experimental Philosophy in 1822. He also became superintendent of the new university observatory: he therefore devoted his last years mainly to astronomy.[5]

Woodhouse's early researches were concentrated on the pure calculus. In (1801b) he dealt with 'Viviani's problem' on the squaring of portions of the surface of a sphere. This problem became famous among the British; it was treated for instance by Brinkley (1802b). Another important area of research encompassed elliptic integrals: Woodhouse (1804) summarizes without any claim of originality the researches carried out into this subject by Euler, Legendre, Lagrange, Landen, Ivory, Wallace and Brinkley. Woodhouse shows a very extensive knowledge of the literature, even though he is not aware of Legendre (1794) which was to be translated in the *Mathematical Repository* (see chapter 8, section 8.5).

Beside these two minor papers Woodhouse contributed to the *Philosophical Transactions* with (1801a) and (1802) on the nature of mathematics and the foundations of the calculus. Later, in *Principles of Analytical Calculation* (1803), Woodhouse developed a systematic attempt to provide a foundation for the calculus which exerted a great influence in

Great Britain. Indeed, in the years 1801–3 he launched a programme of research which was still alive in the 1840s.

Woodhouse (1801a) is concerned with the legitimacy of the use of complex numbers. Even though handling complex numbers was an important practice in eighteenth-century mathematics, the status of the 'roots of negative quantities' was still uncertain. This does not mean that there was consensus on this matter. In particular, the situation in Great Britain was a varied one. We have already met Playfair (1779) in which the geometrical analogy between the circle and the hyperbola was used to explain the meaningfulness of the imaginaries. Extremists such as William Frend and Francis Maseres tried to build up a theory of algebraic equations confined to real solutions. On the other hand, mathematicians such as Simpson, Landen and Waring employed complex numbers in their researches. Woodhouse in his (1801a) was therefore touching on a controversial topic. We would not need to concentrate on this aspect of Woodhouse's work had it not had a relevant influence on his conception of the calculus of fluxions.

According to Woodhouse the operations on imaginaries can be justified via a formal use of infinite series. If the expansion of e^x, when x is a real number, is extended for xi, then:

$$e^{xi} = 1 + xi - x^2/2! - x^{3i}/3! + x^4/4! + \ldots.$$

This extension of a formula which holds for real numbers to complex numbers is typical for Woodhouse of the development of mathematics. For instance he writes:

But nothing can be affirmed concerning the product of $(a+bi)$ and $(c+di)$ [...]; and all that can be meant by the form $(a+bi)(c+di)$ is that, the characters are to be combined after the same manner that the signs of quantity are; so that $(a+bi)$ $(c+di)$ and $ac+adi+cbi-bd$ are two forms equivalent to each other, not proved equivalent, but put so, by extending the rule demonstrated for the signs of real quantities. (Woodhouse (1801a), p. 93)

A similar substitution for $-xi$ in e^x and a termwise subtraction yields the result:

$$\frac{(e^{xi}-e^{-xi})}{2i} = x - x^3/3! + x^5/5! - \ldots = \sin(x).$$

This is just one example of the way in which, according to Woodhouse, all the theorems in mathematics should be treated: they should be reduced in terms of algebraical operations with infinite series. All the mathematical functions such as $\sin(x)$, $\exp(-xi)$, etc., rather than having a geometrical or a mechanical meaning, are just abbreviations of their power series expansions. The calculus is nothing other than an algebraical handling of infinite series. Some of the operations of the calculus can be interpreted in geometry or mechanics, some cannot; but this is incidental and does not concern the mathematician. Hence, for Woodhouse,

By the strange way of determining the meaning and value of analytical expressions from geometrical considerations, it should seem, as if certain curves were believed to have an existence independent of arbitrary appointment. (Woodhouse (1802), p. 92n)

At this stage what Woodhouse needs is a proof that every function has a power expansion, which is the true object of algebraical manipulations. He thought that a proof was feasible in which only algebra was employed. From this point of view the calculus would have received, as in John Landen's *Residual Analysis* (1764), an algebraical foundation. Woodhouse attempted this grandiose project in his *Principles of Analytical Calculation* (1803). The novelty of this work, which lies in the intent to supersede the fluxional tradition, was evident even in the notation. Woodhouse employed the differential notation and also Arbogast's D for the derivative operator.

Lagrange in his (1797) had already attempted an algebraical proof of Taylor's theorem. This work was known to Woodhouse, who reviewed it in (1799),[6] and most probably stimulated not only the *Principles of Analytical Calculation* (1803), but also his papers on complex numbers (1801a) and the nature of mathematics (1802). Lagrange assumed that $f(x+i) = f(x)+iP$, where P is finite when $i = 0$. P is a function of x and i and therefore we can write $P = p+iQ$, where p is the value of P when $i = 0$. By reiterating this process Lagrange obtained:

$$f(x+i) = f(x)+ip+i^2q+ \ldots.$$

He then defined p as the derivative of $f(x)$, while the integral was defined as the inverse of the derivative.

We can end our analysis of Lagrange (1797) here since Woodhouse already had a good objection: the definition of p does not avoid a limit process:

$$p = f'(x) = P \text{ when } i = 0, \text{ where } P = (f(x+i)-f(x))/i.$$

However, Woodhouse did not completely abandon Lagrange's programme. He agreed with Lagrange in considering both the use of differentials and that of limits as ungrounded. In fact, Woodhouse praised Berkeley's criticisms of the calculus, and in the preface to his (1803) he endorsed the arguments against Newton and Leibniz contained in *The Analyst* (1734):

If, for the purpose of habituating the mind to just reasoning, (and mental discipline is all the good the generality of students derive from the mathematics) I were to recommend a book, it should be the *Analyst*. Even those, who still regard the doctrine of fluxions as clearly and firmly established by their immortal inventor, may read it, not unprofitably, since, if it does not prove the cure of prejudice, it will be at least the punishment. (Woodhouse (1803), p. xviii)

Woodhouse's programme therefore coincided with that of Lagrange in many points. He thought that the concepts of motion, limit and infinitesimal were to be avoided, and he believed that the calculus rested upon an algebraical demonstration of power series expansions. But Lagrange was too hasty in his generalizations: according to Woodhouse one had to prove *algebraically* that each function to be employed in the calculus had a Taylor expansion. Once this had been done, one could extend the Taylorian expansions to complex variables and obtain all the results of the calculus by summing and multiplying infinite series.

The problem of convergence was of secondary importance according to Woodhouse: the arithmetical interpretability of series as summations was just a particular application of the 'analytical calculus'. Series had to be considered abstractly as symbolical expressions which exhibit the formal properties of functions. We have seen an example of this procedure in Woodhouse's papers on imaginary quantities. *The Principles of Analytical Calculation* (1803) is a collection of examples in which the following steps occur in succession:

(1)   prove that the functions (e.g. $\sin(x)$, a^x) have a Taylor expansion,
(2)   extend the Taylor expansion to complex values and to values which do not belong to the disc of convergence,
(3)   manipulate algebraically the series,
(4)   as an illustration interpret, if possible, the algebraical results in arithmetic, mechanics or geometry.

As we will see (section 9.5), Woodhouse's programme had a great influence in Cambridge and in Great Britain in general. For instance, many of George Peacock's ideas can be traced back to Woodhouse (1803).[7] Woodhouse himself recognized his indebtedness to Lagrange (1797) and Arbogast (1800). Through him the British came in contact with the Lagrangian algebraical school. As a matter of fact, this school was declining on the continent and it is perhaps unfortunate that so many British mathematicians followed Woodhouse in imitating the Lagrangian school. An attempt to reform the fluxional calculus led once again to a condition of isolation. But why was the Lagrangian school chosen from among the various competing continental schools? It is probable that for British mathematicians this was the easiest step to take. In their eighteenth-century tradition series played a prominent role: Newton's calculus was a calculus of 'series and fluxions'. Banishing the kinematical calculus of fluxions left one with a 'calculus of series' independent of the ideas of motion, limit and infinitesimal moment. For the fluxionists 'pure analytics' coincided with the use of infinite series. When fluxionists abandoned geometrical proofs it was in order to work with power expansions. Power series were used in the study of extremals, curvatures and so on, as well as in the 'integration' of fluxional equations. Hutton's remark is typical of this tradition:

Whenever, in analysis, we arrive to a complex function or expression, either fractional or transcendental; it is usual to convert it into a convenient series, to which the remaining calculus may be more easily applied. [...] If, therefore, we only so far change the received notion of a sum as to say, that the sum of any series, is the finite expression by the evolution of which that series may be produced, all the difficulties [...] vanish of themselves. For, first, that expression by whose evolution a converging series is produced, exhibits at the same time the sum, in the common acception of the term; neither, if the series should be divergent, could the investigation be deemed at all more absurd, or less proper, namely, the searching out a finite expression which, being evolved according to the rules of algebra, shall produce that series. And since that expression may be substituted in the calculation instead of this series, there can be no doubt that it is equal to it. (Hutton (1786), p. 173)

Here Hutton is simply reporting an idea already expressed by Euler. A divergent series, even though it has no sum, can be taken as algebraically 'equal' to a certain function. According to Hutton the algebraical use of series was a permissible technique of 'analysis'. Furthermore, power series were understood as 'infinite' polynomials: their study was part of algebra.

It is with the same methodology that Woodhouse affirmed that convergence is required only in the last stage of calculation, i.e. when the algebraical results are interpreted as numerical results:

the convergency of series is only to be considered, at the end of the calculation, it is of no import to know, whether $(a+x)^m$ converges or not: for analytically, it is the evolution and the law of formation of the coefficients, that it is necessary to know. (Woodhouse (1803), p. 162)

Adopting the Lagrangian foundation of the calculus was therefore a way of systematizing the formal use of series which was already operating in the eighteenth-century calculus. This aspect of the Lagrangian methodology was particularly consonant with the British fluxional tradition.

Woodhouse's reform of British mathematics continued with two textbooks: (1809) is a textbook on trigonometry in which Woodhouse attempted to teach the subject analytically and not geometrically as was usually the case in Britain. This textbook became very popular and was widely read. Woodhouse (1810) is an historical presentation of the calculus of variations from Johann Bernoulli to Lagrange. This text was extremely important because, as we know, the calculus of variations was almost entirely unknown in Great Britain. Woodhouse also wrote three texts on astronomy. Of particular importance was Woodhouse (1818) on physical astronomy in which the main mathematical techniques of Laplace were presented. So after an early period in which he developed a personal view on the foundations of the calculus, Woodhouse devoted himself to the introduction into Britain of Eulerian analytical trigonometry, the Lagrangian calculus of variations and Laplacian physical astronomy. He deserves to be remembered as one of the first and most influential reformers of British mathematics.

9.3 Ireland in the eighteenth century

With the exception of George Berkeley (and his scarcely original opponents Jacob Walton and John Hanna), no interest in the calculus emerged in Ireland during the eighteenth century. Despite this, the influence of Irish science in Great Britain was not negligible. Hugh Hamilton (1729-1805), Erasmus Smith Professor of Natural Philosophy, wrote an excellent treatise on conics (1758), while Richard Helsham's (1682-1738) (who was Donegal lecturer in mathematics) lectures on natural philosophy (1739), read in Trinity College (Dublin), were still widely used in the 1800s.

An effort to revive the study of mathematics was made in the 1760s by one Joseph Fenn, who described himself in the title-page of his book (1769, 1772) 'Professor of Philosophy in the University of Nantes'. In 1768 he was employed by the Dublin Philosophical Society to teach a course which included 'mathematics, the Physical System of the World, the Moral System of the World, Military Art, Merchantile Art, Naval Art, Mechanic Art' (Fenn (1769, 1772), I, p. iii).[8] Fenn wrote two volumes for his course 'to be given in the Drawing-School' in Dublin. In the preface he emphasized the importance of the 'analytic art' as cultivated by the 'first mathematicians in Europe'. The study plan in Fenn's preface indicated 'Sublime Geometry' which 'comprehends the inverse method of Fluxions, and its application to the Quadrature of Curves, the Cubing of Solids, &c.' with the improvements of 'Cotes, Bernoully, Euler, Clairaut, d'Alembert, M'Laurin, Simpson, Fontaine' (Fenn (1769, 1772), I; pp. xiff). This was an outstanding project for the Great Britain of the 1760s: a project which was not fulfilled in the slightest. Fenn's two volumes consisted of an introduction to Euclid's *Elements* and the 'Elements of numeral and specious arithmetic'; in this last section some space was given to Newton's method of reversion of series.

In 1782 the Royal Irish Academy was founded, and in 1787 a series of volumes of *Transactions* was launched. Chairs of Mathematics and Astronomy were founded in the University of Dublin in 1762 and 1774, respectively. These Chairs were occupied, however, by third-rank mathematicians such as Richard Murray (d.1799), William Magee (1766–1831) and Henry Ussher (d.1790). The first six volumes of the *Transactions of the Royal Irish Academy* (1787–97) reflect the very poor level of Irish mathematics in the second half of the century. A few papers by Matthew Young (1750–1800) on series and equations could have been easily written a century before. Young, however, was mainly interested in sound (1784). William Hales (1747–1831), Professor of Oriental Languages and Divinity at Dublin University, wrote a 'conservative' vindication of the Newtonian fluxional calculus in (1800), a work concerned with foundations. Hales also wrote on sound (1778), the motion of planets (1782) and equations (1784). Other texts on optics and natural philosophy (see, e.g., Stack (1793), Miller (1799) and Young (1800)) indicate the interest of the Irish in science, but there was no use at all of mathematics.

9.4 The reform of mathematics at the University of Dublin

A great change occurred in 1790 with the election of John Brinkley (1763–1835) to the Andrews Chair of Astronomy at Trinity College, Dublin.[9] Brinkley was born in England and was educated at Cambridge where he graduated B.A. in 1788 as first Smith's prizeman. Like Robert Woodhouse, Brinkley was influenced by Waring and was interested in astronomy. In the year 1787–8 he worked as the assistant of Nevil Maskelyne. Later in 1792 he became first Astronomer Royal of Ireland. The strongest influence on Brinkley was to be exerted by Laplace: as in the cases of Playfair and Ivory, it was the *Mécanique Céleste* which stimulated Brinkley's approach to the continental calculus.

Brinkley's researches appeared almost exclusively in the volumes of the *Transactions of the Royal Irish Academy*. He was virtually the only one who contributed on mathematics up to the mid-1820s when William Rowan Hamilton began his career. Brinkley (1800a) and (1800b) were inspired by Waring's *Proprietates Algebraicarum Curvarum* (1772); they dealt with trigonometric formulas and Cotes's theorem. The paper (1800c) was based on Taylor's theorem. Brinkley claimed that he had anticipated Arbogast (1800) (one of Woodhouse's sources in (1803)), a treatise in which the calculus was based on Taylorian expansions. In a later paper (1807) he dealt with the formula:

$$\Delta_h^n u = (e^{h\frac{d}{dx}} - 1)^n u,$$

which had been used by Lagrange, Laplace and Arbogast, expressing it in Newtonian notation as:

$$\Delta_h^n u = (e^{h\frac{\dot{u}}{\dot{x}}} - 1)^n.$$

As we will see below, this trend of research was attributed great importance by the Cambridge reformers of the Analytical Society. Brinkley (1802a) was a review of the principal methods of dealing with the motion of the apsides. Brinkley criticized Walmesley's defence of Newton's method of neglecting the component of the disturbing force of the Sun perpendicular to the Moon–Earth radius vector, and reached an appreciation of the more elaborate work of Clairaut (1752). No great originality is to be found either in (1803a), which is a review of the various solutions of Kepler's problem. The great interest of the British (e.g. Landen, Wallace, Ivory) in elliptic integrals was the motivation behind (1803b), but here, despite his awareness of Lagrange and Legendre, whose works are cited, Brinkley does not contribute any new results.

None of these early papers shows any great originality, but rather a good knowledge of continental literature. They are written in fluxional notation, but the conception is completely continental. Brinkley sometimes embraces a Lagrangian operational view of the calculus, but more often he keeps close to a differential approach. In a later period Brinkley devoted his attention to improving details of Laplace's *Mécanique*. In (1820a) he modified Laplace's method of approximating the mean motion of lunar perigee, while in (1820c) he simplified Laplace's method of calculating cometary paths.

As a teacher and reformer of mathematics Brinkley exerted a certain influence in the University, even though he resided outside Dublin at the Dunsink Observatory. As Playfair had done in his *Outlines of Natural Philosophy* (1812–14), in *Elements of Astronomy* (1813) Brinkley introduced the students to Laplace's theory of planetary motions. In doing so, Brinkley certainly improved the level of teaching in Dublin, even though he avoided mathematical technicalities. His efforts were later backed by Bartholomew Lloyd (1772–1837), who succeeded Magee in the Chair of Mathematics. It is Lloyd that introduced Lacroix's treatise (1802) and Poisson's textbook on mechanics (1811) into the curriculum at Trinity College. Therefore, while it was in the 1790s with Brinkley that Dublin

students began reading about physical astronomy, the *differential* calculus
arrived with Lloyd in the 1810s. It is important to note that in Ireland
there was no passage from the study of fluxions to the study of
differentials; instead Dublin University went from the absence of calculus
to the differential calculus.[10]

The major interests in Dublin were, however, not in the calculus:
treatises were written on astronomy, analytic geometry and mechanics.
The differential calculus was taught especially in the study of mechanics.
For instance, Thomas Robinson's *System of Mechanics* (1820), a treatise
intended for the students at Dublin, was based on Poisson (1811). Each
chapter was followed by a mathematical appendix for the more interested
students. Robinson's appendixes are written in differential notation; the
calculus of variations is sometimes given in a Lagrangian style, while
d'Alembert's principle is often used in the treatment of the dynamics of
rigid bodies. We should note that in England the first non-Newtonian
presentations of mechanics were to be found in a translation of two
textbooks by Venturoli ((1822) and (1823)), in which the principle of
virtual velocities was employed, and in the first editions of William
Whewell's *Elementary Treatise on Mechanics* (1819) and *Treatise on
Dynamics* (1823). Chapter IX of (1819) was devoted to virtual velocities,
while in Book III of (1823) Whewell employed 'd'Alembert's principle' in
the study of rigid bodies, and in Appendix H he compared it with other
principles in mechanics. Later on Whewell returned to Newton's *Principia*
and always declared his opposition to the continental approaches to
mechanics.

Robinson (1820) is representative of the way in which the reform of
mathematics took place in Dublin. The subjects of interest were Laplacian
astronomy, Monge's geometry and Poisson's mechanics; mathematics
was studied within this applied context. The fruits of this reform came in
the next generation with Humphrey Lloyd, MacCullagh and William
Rowan Hamilton.[11]

An 'Irish' treatise on the differential and integral calculus appeared only
in 1825. The author, Dyonisius Lardner (1793–1859), was educated at
Trinity College, Dublin. In 1827 he was appointed to the Chair of Natural
Philosophy and Astronomy at the newly founded University of London
(which was to become University College). Later he became a prolific
scientific writer and a successful publisher. His *Elementary Treatise on the
Differential and Integral Calculus* (1825) was divided into four parts:
differential calculus, integral calculus, calculus of variations and finite
difference equations. Lardner (1825) can be seen chronologically as the
third 'continental' treatise published with the aim of reforming the

calculus in Great Britain. The first, Wallace (1815), came from the Military School of Sandhurst; the second, Lacroix (1816), was the translation of the second edition of Lacroix's *Traité Elémentaire* (1802). This translation was prepared at Cambridge by a group of young reformers to whom we will now turn our attention.

9.5 The Analytical Society

When, in 1812, the Analytical Society of Cambridge was established, the time was ripe for a shift towards the continental calculus and mechanics. French work in the field of engineering was followed by the Woolwich men. Legendre's elliptic integrals were considered by Wallace and Ivory in the *Transactions of the Royal Society of Edinburgh* and the *Mathematical Repository*, and by Hellins and Woodhouse in the *Philosophical Transactions*. Lagrange's and Arbogast's algebraical foundations of the calculus were adopted by Brinkley, Woodhouse, Bonnycastle and Spence.[12] Most of all, Laplace's *Mécanique* provided a stimulus to learn French mathematics. The works of Brinkley and Ivory were almost entirely devoted to improving aspects of Laplace's masterpiece. John Toplis (1805) and Playfair (1808) indicated Laplace as the model to be followed and regretted the gap which separated French and British mathematical research and teaching. French science in general was colonizing British scientific journals and British encyclopaedic works.

Even though Cambridge educated men such as Brinkley and Woodhouse had participated with enthusiasm in the reforming of British mathematics, little effort was made to change the teaching at Cambridge. Bright students were therefore in a mood of protest. A group of undergraduates, among whom were most notably George Peacock (1791–1858), Charles Babbage (1791–1871) and John Herschel (1792–1871), founded in 1812 the 'Analytical Society'. Its objective was to propagate the heresy of 'pure d-ism against the Dot-age of the University'.[13] A project was set up to translate the second edition of Lacroix's short treatise on the calculus (1802): the aim of the Analytical Society's members was clearly that of changing the kind of education provided at Cambridge. However, the Analytical Society collapsed around 1814, having produced only a volume of *Memoirs* ([Babbage and Herschel] (1813)). The opposition was very strong. The pedagogical ideas which were prevalent in the early nineteenth century (and still existing in Victorian Britain) did not allow them to 'reduce' a university to a centre for research in pure mathematics. Robert Woodhouse, who confined his interest in Lagrangian mathematics to his private research, did not provoke any scandal: he did not question the purpose of Cambridge education. The Analytical Society's programme, on the other hand, caused a loud outcry. Lacroix's treatise was intended for the students at Cambridge, as were the volumes of exercises on differential and integral calculus, functional equations and finite differences, Babbage, Herschel and Peacock (1820), which were published in 1820. After the collapse of the Analytical Society these works were published through the efforts of Babbage, Herschel and Peacock. After 1820 only Peacock remained at Cambridge; he developed an approach to algebra which had many points of contact with Woodhouse's foundation of the calculus. Herschel continued his father's work on astronomy; Babbage was engaged in work on his difference and analytical engines. The reform of mathematical education at Cambridge went ahead in a more moderate and tortuous form than had been wished by the group of undergraduates who had got together in 1812.[14]

The struggle for the reform of mathematical education at Cambridge has been mistakenly viewed as a process in which British mathematics was successfully reformed. Historians largely based this view on recollections of former members of the Analytical Society who liked to describe themselves as the originators of interest in continental mathematics and the revival of research in the first half of the nineteenth century. This is clearly false. They were anticipated by Woodhouse at Cambridge, and as we have seen there were several other centres of reform as equally important as Cambridge which should be considered.

However, Babbage, Herschel and Peacock were of great importance for the early nineteenth-century British calculus. Their contributions did not consist, as is usually maintained, in the introduction of differential notation into Great Britain. The shift to differential notation was already underway in the first years of the century. The contributions of Babbage, Herschel and Peacock are twofold.

In the first place, they reinforced the acceptance of the Lagrangian algebraical foundation of the calculus, already espoused by Brinkley, Woodhouse, Spence and Bonnycastle. (Lacroix's treatise (1802) was not in fact a Lagrangian treatise, but rather a mélange of methods in which limits were prominent; but the translators added Lagrangian notes to it.) Accepting the Lagrangian view on the nature of the calculus had many consequences on the way in which research was done, especially in the case of infinite series. The difficulties that many British Lagrangians had in understanding the importance of Cauchy's treatment of series depended on the way in which they understood the problem of convergence as not belonging to the 'pure' calculus.[15] Furthermore, Cauchy's limit-based calculus was felt by the British Lagrangians to be a revival of the fluxionists' theory of prime and ultimate ratios. The slow acceptance of Cauchy's calculus in Great Britain had its roots in the domination of the Lagrangian tradition promulgated by Babbage, Herschel and Peacock.

But the second and most important contribution from the Analytical Society's members was that they initiated a trend of research which characterized much of British mathematics up to Cayley and Boole, The *Memoirs of the Analytical Society* were centred on the calculus of operators and on functional equations.

The calculus of operators dealt with the algebraical properties of the symbols of derivative and integral, and the related symbols of finite difference and summation. From this study it was possible to develop symbolic methods of integration of differential and difference equations. For instance, a differential equation was written, 'separating the symbols of operation from that of quantity' as $f(D)y = X$. Then $f(D)$ was manipulated algebraically in order to find the inverse, $f^{-1}(D)$. In the simple case $(D^2 - (a+b)D + ab)y = X$, treating $f(D)$ as a polynomial in D, it is possible to obtain:

$$y = \frac{((D-a)^{-1} - (D-b)^{-1})X}{a-b}.$$

In the case of finite difference equations, a significant role was played by the symbolic equivalence:

$$\Delta_h^n = (e^{h\frac{d}{dx}} - 1)^n.$$

This last result, due to Lagrange, was treated in fluxional notation in Brinkley (1807). As a matter of fact, the application of the calculus of operators to the theory of integration originated mainly from Lagrange. His research was continued by a group of French mathematicians which included L. F. A. Arbogast, F. J. Servois, J. F. Français and B. Brisson. Nevertheless, on the continent the Lagrangian school never played a prominent role. In Great Britain, on the contrary, the introduction of operational methods with Brinkley (1807), Spence (1809), the Analytical Society's *Memoirs* and the notes and appendixes to Lacroix (1816) launched a programme of research which continued up until the 1840s.[16]

In addition, use of the calculus of functions started in Great Britain with the Analytical Society's members, especially Babbage. The problem of recognizing the form of the arbitrary functions which occur in the integration of partial differential equations was the motivation for developing a theory of functional equations. Functional equations considered in this early period had the quite general form:

$$F(x,\psi(f_1(x)), \psi(f_2(x)), \ldots, \psi(f_n(k))) = 0,$$

where ψ is an unknown. These equations were reduced to finite difference equations for $f_m(x) = x + mh$. So the calculus of functions was viewed as an extension of the calculus of finite difference operators. The importance given to this theory is demonstrated by the inclusion in the *Encyclopaedia Metropolitana* of a lengthy essay on 'functional equations' by the young Augustus De Morgan (1836).

From the 1730s to the 1760s there was a great production of Lagrangian mathematics in England. I have already pointed out, in the case of Woodhouse, that the Lagrangian foundation of the calculus on series was, to a certain extent, compatible with the fluxional tradition: to banish kinematics from the method of series and fluxions meant to be left with only series. Undoubtedly, there is a great epistemological difference between the two methods, but at the level of mathematical practice the algebraical manipulations with series of the Lagrangians were, to a certain extent, similar to the methods employed by the analytical fluxionists, such as Simpson, Landen and Waring. But the Analytical Society's members brought into Britain not only the Lagrangian use of series, but especially operational methods and functional equations. These techniques became extremely popular in Britain, as the reader of the *Cambridge Mathematical Journal* knows very well. Hundreds of papers on solving differential equations by operators were published. The reform of mathematics which took place in Cambridge had the effect of diverting the attention of British mathematicians (or should we still call them 'philomaths'?) from fluxions to Lagrangian methods. The main reason for the success of these techniques is that they were easy to learn and offered immense possibilities of dull proliferation. Furthermore, the followers of Babbage, Herschel and Peacock (who, by the way, very soon realized the sterility of their infatuation with Lagrangian analytics), considered the calculus of operators as a 'new continental method', and they were clearly excited to participate in this Renaissance.

The researches of the British Lagrangian school on the calculus of operators and the calculus of functions were the origin of important contributions to algebra and logic, such as Peacock's 'pure algebra', and De Morgan's and Boole's algebras of logic. But the predominance of the algebraical approach to the calculus had its own drawback: it did not allow many British mathematicians influenced by the Analytical Society to appreciate the importance of Cauchy's rigorization of the calculus, which was motivated by the desire to avoid the 'generalities of algebra'. The shift from the fluxional calculus to the Lagrangian calculus, which marks the definitive death of the Newtonian tradition, once again left the British isolated.

CONCLUSION

HAS MY research been successful in refuting the accepted views on the crisis of the Newtonian calculus? None of these views corresponds to the image we obtain from a close scrutiny of the fluxional texts. Nevertheless, the label 'dotage' still attaches to the treatises on fluxions we have encountered. A crisis did occur, but it set in later than is usually thought.

During the first four decades of the eighteenth century the calculus of fluxions was developed by men such as Cotes, Taylor, Stirling and Maclaurin. With their work they made original and important contributions to the fields of integration, series and applied mathematics. The British began losing ground in the 1740s. In the mid-eighteenth century only Simpson and Landen achieved new results. Very soon many British mathematicians realized the gap which separated them from the continentals and several attempts were made to change this situation of isolation. We have seen how slow and complex was this process of reform, which began with Simpson, Landen and Waring, and was continued especially by Playfair, Ivory, Wallace, Woodhouse, Brinkley and the Analytical Society's fellows.

The era of the Newtonian calculus cannot be simply described as a period of decline. It was a period of the history of British mathematics which began with successes, suffered a period of crisis, and ended with serious attempts to reform.

In what exactly consisted the crisis? At the beginning of the century British mathematics was in close contact, and sometimes in bitter competition, with the rest of Europe: but by the middle of the century it was almost completely separated from the continent. The works of continental mathematicians were not understood in Britain, while the works of the British aroused little interest on the continent.

We have to realize that as the century progressed the continental calculus underwent deep changes. A calculus of variable quantities was

139

replaced by a calculus of multivariate functions. This resulted in new perspectives and new problems. First of all it was necessary to codify the basic rules of the direct method of partial differentiation, such as $\partial/\partial x \partial y = \partial/\partial y \partial x$ (a relation which was understood as universally true). Secondly, the inverse method of integration produced solutions of partial differential equations, the investigation of multiple integrals and contour and surface integration. The development of these tools had enormous consequences on the way in which mechanics could be mathematized. In turn mechanics provided a stimulus to perfect the new methods of the calculus of multivariate functions and the calculus of variations.

The British almost completely missed these new developments. This was very sad, since several aspects of Newtonian mathematics were well adaptable for the change. Indeed the concept of fluent was closer to that of function than the concept of differential; but the concept of function appeared only sporadically, in the works of Landen and Spence. Furthermore, partial differentiation could be achieved in the context of the Newtonian calculus, and here some evidence is available. Techniques analogous to partial differentiation (in the context of total differentials and multiple integration) were present in the works of some of the fluxionists. When fluxionists considered an expression involving more fluent quantities they read it as a function of time:

$$F(x(t), y(t), z(t), \dots).$$

Taking the fluxion meant taking the total differential, and, of course, expressions equivalent to partial differentials occurred in it. Several fluxionists used to begin by stating that 'all the fluents except x' were to be considered as constants: then one could take the fluxion by implicit differentiation as a function of time. This process was repeated for y, z, etc.: this is how the terms of the total differential were obtained. However, there was no specific symbol for partial differentiation, such as Euler's bracketed (df/dy), and the properties of partial differentiation were not studied. The fluxional equivalent of partial differentiation was not recognized as a mathematical concept whose properties could be studied, but it was understood as a tool in the algorithm employed for calculating total differentials.

Also the fluxional equivalent of multiple integration was employed by the British. It was quite natural to take, for instance, $F(x(t), y(t))$ first as an expression in which only y is flowing. The first fluent of it, $\int F dy$ in Leibnizian notation, could then be considered as an expression in which only x is flowing, and the fluent was calculated again: $\int\int F dy dx$. As such, partial differentiation and multiple integration had been used since the times of Newton in calculating total differentials and volume integrals. However, contour and surface integration, partial differential equations and the calculus of variations were never understood by the British in the eighteenth century.

Why did the British stop there? There is not an easy answer to this question. Part of the difficulty in finding an answer probably lies in the way in which eighteenth-century calculus is viewed. According to the standard view, after the dramatic 'discovery' of the calculus all that the eighteenth-century mathematicians could do was to extend the boundaries of the calculus, in a haphazard process of colonizing new areas and new formulas which were unscrupulously conquered and added to the body of mathematical knowledge. Only in the nineteenth century would there be another foundational reworking, with Cauchy's rigorization of the calculus.

However, it seems to me that the eighteenth century was not a period of Kuhnian normal science (to use a fashionable terminology): there was no simple puzzle-solving on the part of Euler, for instance. During this century the calculus was transformed: in a way a new calculus was created.

The difficulties that many gifted British mathematicians had in understanding this new calculus should therefore be understood in this perspective. The problems and misunderstandings which emerged in the process of translating the continental calculus in the language of fluxions show us the extent of our underestimation of the novelty of eighteenth-century calculus.

Finally, let us consider the principal characteristics of the reforms of the calculus.

A new consciousness of the importance of reforming the Newtonian tradition at the levels of both research and teaching came from France at the turn of the century. The *Mécanique Céleste* began to be published in 1799. Laplace had achieved outstanding results and was seen as a second Newton. His researches were received in Great Britain as a solution of problems which were left to be answered in the *Principia* by Newton. Reading Laplace, and hopefully understanding it, became imperative.

In the reform which followed were involved several schools and mathematicians. We classified them in the four groups geographically situated in Scotland, the military schools at Woolwich and Sandhurst, Dublin and Cambridge.

In Scotland and in the Royal Military College at Sandhurst the concern with Laplacian astronomy was at the level of research and the teaching was little changed. A radical reform of the teaching motivated the Dublin and Cambridge reformers. The Dublin group insisted more on the teaching of applied mathematics, whereas the Cambridge group was definitely purist–algebraist. A distinction should be made also at the level of research interests: while in Dublin the interest turned towards Laplace, in Cambridge the algebraical calculus of Lagrange came in.

Although the Cambridge reform of the *teaching* was almost a complete failure, the Analytical Society's advocacy of the Lagrangian calculus of operators and the calculus of functions spread all over Britain as a tradition for *research*. It is again unfortunate that many of the British opted for Lagrange's approach, for on the continent itself it was becoming a bit old fashioned, while a new school was gradually emerging with Cauchy's rigorization of the calculus. In following Lagrange, many British mathematicians rather isolated themselves from the continent. How this isolation brought new and unsuspected discoveries in the algebra of logic and other related fields would be the subject for another book.

APPENDIX A

TABLE OF CONTENTS OF FLUXIONARY TEXTBOOKS

A.1 Table of contents of William Emerson's *The Doctrine of Fluxions* (1743) (2nd edn., 1757)

Preface

The Doctrine of Fluxions
Postulatum
Definitions
Notation
Axioms

Section I The Fundamental Principles and Operations of Fluxions

Prop.I. The fluxion of any fluent or generated quantity is equal to the sum of the fluxions of all the roots or sides, each multiply'd continually by the index of its power, and by the given fluent when divided by the said root or side

Prop.II If two fluents or variable quantities be equal to each other, or in a given ratio; their fluxions will be equal, or in the same given ratio. And if two flowing quantities be equal or in a given ratio, their contemporary fluents will be equal, or in the same given ratio

Prop.III To find the fluxions of quantities Prop. IV–VIII several theorems for finding the fluents of quantities from other given fluents

Prop.IX To transform fluxional quantities into others equal to them

Prop.X To find the fluents of quantities by infinite series

Prop.XI To find fluents by the table

Prop.XII To correct the fluent of a given fluxion

Prop.XIII To investigate a problem by the method of fluxions

Section II The investigation and solution of some of the most general
and useful problems in the mathematics

Prob.I To find the maxima and minima of quantities

Prob.II To find the logarithms of numbers

Prob.III To draw tangents to curves

Prob.IV To find the points of contrary flexion of a curve

Prob.V To find the radius of curvature of a curve

Prob.VI To find the variation of curvature

Prob.VII To find the nature of a curve, by whose evolution a given
curve is described

Prob.VIII To find the lengths of curve lines

Prob.IX To transform spirals into geometrical curves, or geometrical
curves into spirals of equal lengths

Prob.X To find the quadrature of curves

Prob.XI To find curves that are quadrable

Prob.XII To find curves whose areas shall have any assigned relation to
the area of a given curve

Prob.XIII To find the surfaces of solids

Prob.XIV To find the solidity of bodies

Prob.XV To find catacaustick curves

Prob.XVI To find diacaustick curves

Prob.XVII To find the centre of gravity

Prob.XVIII To find the centre of percussion and oscillation

Prob.XIX To find the law of centripetal force for a given curve

Prob.XX The nature of an arch being given, to find the height of the
wall upon any part of it

Prob.XXI The nature of a concave vault or cupole being given, to find
the height of the wall in any point

Prob.XXII To find the resistance of a body moving in a fluid

Prob.XXIII [2nd edn.] To find the centre of gyration

Prob.XXIV [2nd edn.] To find the strength of a piece of timber of any
figure

Section III The investigation of physical problems

Prob.I To find the curve which a flexible line is put into by the force of
the wind or any fluid

Prob.I [2nd edn.] To find the fluxions of the times, velocities, and spaces
described by bodies in motion; being acted upon by any accelerating
force: a universal problem

Prob.II To find the motion of a musical string

Prob.III To find the velocity of a projectile moving in a given curve

Prob.IV To find the velocity of a body, falling towards the Earth, according to any law of gravity

Prob.V To find the time of a body's descent towards the Earth according to any law of gravity

Prob.VI From the velocity and direction of a projectile, and the law of centripetal force; to find the velocities, times, and angles of revolution

Prob.VII To find the time of a pendulum's vibrating in the arch of a cycloid

Prob.VIII To find the force wherewith a corpuscle is attracted towards the plane of a circle, according to any law of centripetal force

Prob.IX To find the force wherewith an infinite solid attracts a corpuscle, when the law of attraction is inversely as some power of the distance greater than 1

Prob.X To find the force wherewith a sphere attracts a corpuscle, when the force of every particle is reciprocally as the square of the distance

Prob.XI To find the force wherewith a spheroid attracts a corpuscle placed at the pole

Prob.XII To find the motion of light passing through a refracting medium

Prob.XIII To find the motion of a globe in a resisting medium

Prob.XIV To find the motion of a globe ascending or descending in a resisting medium

Prob.XV To find the motion of a globe oscillating in a cycloid in a resisting medium

Prob.XVI To find the density of the atmosphere at any height, according to any law of gravity, supposing the density of the air as the compression

Prob.XVII To find the density of the atmosphere at any height, according to any law of gravity, supposing the compression of the air to be as any power of the density

Prob.XVIII To find the polar and equinoctial diameters of the Earth

Prob.XIX [2nd edn.] To find the motion of a projectile in a resisting medium

Prob.XX [2nd edn.] To find the time of ascent and descent of a fluid in the legs of a canal or crooked pipe

Prob.XXI [2nd edn.] To find the weight of a ball falling a given height, which shall break a bar whose strength is given

Prob.XXII [2nd edn.] To find the velocity of the motion propagated through a number of ivory balls, by the impulse of the first upon the second

Prob.XXIII [2nd edn.] To find the velocity of sound

A.2 Table of contents of Thomas Simpson's *The Doctrine and Application of Fluxions* (1750c)

Preface

Part the first

Part the second

Section V The investigation of fluents of rational fractions, of several quantities, according to the forms in Cotes's Harmonia mensurarum

Section VI The manner of investigating fluents, when quantities, and their logarithms, arcs and their sines, &c. are involved together: with other cases of the like nature

Section VII Showing how fluents, found by means of infinite series, are made to converge

Section VIII The use of fluxions in determining the motion of bodies in resisting mediums

Section IX The use of fluxions in determining the attraction of bodies under different forms

Section X Of the application of fluxions to the resolution of such kinds of problems de maximis et minimis, as depend upon a particular curve, whose nature is to be determined

Section XI The resolution of problems of various kinds

A table of hyperbolical logarithms

A.3 Table of contents of John Rowe's *An Introduction to the Doctrine of Fluxions* (1751)

Preface

Part I

Of the principles of fluxions, and of the new notation in algebra
Of finding the fluxion of a given fluent
Of drawing tangents to curves
Of finding maxima and minima
Of finding points of inflexion in curves
Of finding the radius of curvature
Of finding the nature of the evolute of a given involute curve

Part II

Of infinite series
Of finding the fluent of a given fluxion
Of finding the length of a curve line
Of finding the areas of curvilinear spaces
Of finding the convex superficies of solids
Of finding the contents of solids

Part III

Miscellaneous questions, with their incremental and fluxional solutions

APPENDIX B

PRICE LIST OF MATHEMATICAL BOOKS PRINTED FOR JOHN NOURSE

From William Emerson's *Arithmetic of Infinites* London, J. Nourse, 1767.

I *The Elements of Trigonometry*, by W. Emerson. The second edition, with large additions; together with the tables of sines, logarithms, etc. 8vo. 1764. Price 7s.

II *The Elements of Geometry*, by W. Emerson 8vo. 1763. Price 5s.

III *A Treatise of Arithmetic*, by W. Emerson. 8vo. 1763. Price 4s. 6d.

IV *A Treatise of Algebra*, by W. Emerson. 8vo. 1764. Price 7s.

V *Emerson's Navigation*. The second edition. 12mo. 1764. Price 4s.

VI *A new Method of Increments*, by W. Emerson. 8vo. 1763. Price 7s. 6d.

VII *A Treatise of Algebra*, by T. Simpson. The third edition. 8vo. 1766. Price 6s.

VIII *Essays on several curious and useful Subjects in Speculative and Mixed Mathematics*, by T. Simpson. 4to. 1740. Price 6s.

IX *Mathematical Dissertations on a Variety of Physical and Analytical Subjects*, by T. Simpson. 4to. 1743. Price 7s.

X *Miscellaneous Tracts on some Curious and very Interesting Subjects*, by T. Simpson. 4to. 1757. Price 7s.

XI *Trigonometry, Plane and Spherical, with the Construction and Application of Logarithms*, by T. Simpson. The second edition. 8vo. 1765. Price 1s. 6d.

XII *The Doctrine of Annuities and Reversions*, by T. Simpson: 8vo. 1742. Price 3s.

XII *Appendix* to ditto. 8vo. Price 6d.

XIV *The Elements of Geometry*, by T. Simpson. The second edition, with large alterations and additions. 8vo. 1760. Price 5s.

XV *Select Exercises for Young Proficients in the Mathematicks*, by T. Simpson. 8vo. 1752. Price 5s. 6d.

XVI *The Doctrine and Application of Fluxions*, by T. Simpson. 2 vol. 8vo. 1750. Price 12s.

XVII *The Elements of Euclid*, by R. Simson. The second edition, to which is added *The Book of Euclid's Data*. 8vo. 1762. Price 6s.

XVIII *Introduction a L'arithmetique Vulgaire*, 4to. 1751. Price 1s.

XIX *The Elements of Navigation*, by J. Robertson. The second edition. 2 vol. 8vo. 1764. Price 18s.

XX *The Elements of Astronomy*, translated from the French of Mons. De la Caille, by J. Robertson. 8vo. 1750. Price 6s.

XXI *The Military Engineer; or a Treatise on the Attack and Defence of all Kinds of Fortified Places*. In two parts. 8vo. 1759. Price 9s.

XXII *The Accountant; or, The Method of Book-Keeping*. By J. Dodson. 4to. 1650. Price 4s. 6d.

XXIII *The Mathematical Repository*, by J. Dodson. 3 vol. 12mo. 1748–55. Price 12s.

XXIV *An Early Introduction to the Theory and Practice of Mechanics*, by S. Clark. 4to. 1764. Price 6s.

XXV *The British Gauger; or, Trader and Officer's Instructor in the Royal Revenue of Excise and Customs. Part I. Containing the Necessary Rules of Vulgar and Decimal Arithmetic, and the Whole Art of Practical Gauging, both by Pen and Rule; illustrated by a Great Variety of curious and useful Examples. Part II. Containing an Historical and Succinct Account of all the Laws Relating to the Excise: To which are added, Tables of the old and new Duties, Drawbacks, &c. on Beer, Ale, Spirits, Soap, Candles, &c. with a large and copious Index*, by Samuel Clark. 12mo. 1765. Price 5s.

XXVI *The Elements of Fortification*, by J. Muller. The second edition. 8vo. 1756. Price 6s.

XXVII *The Method of Fluxions and Infinite Series, with its Application to the Geometry of Curve Lines, by the Inventor Sir Isaac Newton. Translated from the Author's original Latin. To which is Subjoined, a Perpetual Comment upon the whole Work*, by J. Colson, F.R.S. 4to. 1736. Price 15s.

APPENDIX C

CHAIRS IN THE UNIVERSITIES

C.1 The Gregory's family tree

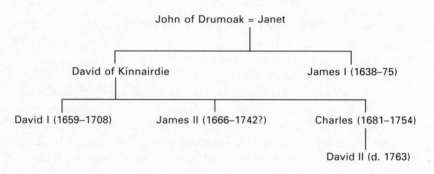

John of Drumoak = Janet

David of Kinnairdie   James I (1638–75)

David I (1659–1708)   James II (1666–1742?)   Charles (1681–1754)

David II (d. 1763)

The Gregorys (Gregories) have been perhaps the most influential family in the history of Scottish universities. From the family tree I filter, as computer theorists would say, only the professors of mathematics. Their careers can be traced in the following appendixes. On the Gregorys see Paul David Lawrence's Ph.D. thesis (1971).

C.2 University of Cambridge

Lucasian Chair of Mathematics

1701 William Whiston lectures as substitute of Newton.
1702 William Whiston appointed.
1710 William Whiston banished for his Arian heresy.
1711 Nicholas Saunderson appointed.
1739 Death of Saunderson.
      John Colson appointed.

1760 Death of Colson
      Edward Waring appointed.
1798 Death of Waring.
      Isaac Milner appointed.
1820 Death of Milner.
      Robert Woodhouse appointed.
1822 Robert Woodhouse moves to the Plumian Chair (see below).

Plumian Chair of Astronomy

1707 Roger Cotes appointed.
1716 Death of Cotes.
      Robert Smith, Cotes's cousin, appointed.
1760 Smith retires.
      Anthony Shepherd appointed.
1796 Death of Anthony Shepherd.
      Samuel Vince appointed.
1821 Death of Vince.
1822 Robert Woodhouse appointed.
1827 Death of Woodhouse.

C.3 University of Oxford

Savilian Chair of Astronomy

1691 David I Gregory* appointed.
1708 Death of Gregory.
1709 John Caswell appointed.
1712 Death of Caswell.
      John Keill appointed.
1721 Death of Keill.
      James Bradley appointed.
1762 Death of Bradley.
1763 Thomas Hornsby appointed.
1810 Death of Hornsby.

Savilian Chair of Geometry

1649 John Wallis appointed.
1703 Death of Wallis.
      Edmond Halley appointed.

* See the Gregory's family tree, p. 150.

1742 Death of Halley.
      Nathaniel Bliss appointed.
1764 Death of Bliss.
1765 Joseph Betts appointed.
1766 Death of Betts.
      John Smith appointed.
1796 John Smith resigns?
1797 Abram Robertson appointed.
1810 Abram Robertson resigns (moves to the Chair of Astronomy)

C.4 University of Edinburgh
Chair of Mathematics

1664 James I Gregory* appointed.
1675 Death of James I Gregory.
1675–83 Chair vacant (John Young, a student, teaches).
1683 David I Gregory* appointed.
1691 David I Gregory resigns (moves to the Chair of Astronomy in Oxford).
1692 James II Gregory* appointed.
1725 Colin Maclaurin appointed joint-Professor (George Campbell competes).
      James Gregory retires from teaching.
1742? Death of James II Gregory.
1746 Death of Maclaurin.
1747 Matthew Stewart appointed.
1772 Matthew Stewart retires from teaching.
      Dougald Stewart, son of Matthew, teaches.
1775 Dougald Stewart appointed joint-Professor.
1785 Death of Matthew Stewart, Dougald Stewart appointed to the Chair of Moral Philosophy.
      Adam Ferguson appointed, John Playfair appointed joint-Professor.
1805 John Playfair appointed to the Chair of Natural Philosophy, John Leslie appointed.

C.5 University of Glasgow
Chair of Mathematics

1691 George Sinclair appointed.
1699 Robert Sinclair appointed (teaching discontinued).

* See the Gregory's family tree, p. 150.

1710 Robert Sinclair resigns.

1711 Robert Simson appointed.

1761 Robert Simson retires from teaching.
James Williamson appointed assistant and successor of Simson.

1768 James Williamson appointed.

1789 James Millar appointed assistant and successor of Williamson.

1795 James Millar appointed.

1832 Death of James Millar.

C.6 University of St Andrews

Chair of Mathematics

1699 James I Gregory* appointed.

1674 James I Gregory moves to the Chair of Mathematics in Edinburgh.

1688 James II* Gregory appointed.

1691 James II Gregory moves to the Chair of Mathematics in Edinburgh.

1707 Charles Gregory* appointed.

1739 Charles Gregory retires.
David II Gregory* appointed.

1763 Death of David II Gregory.

1765 Nicolos Vilant appointed.

1807 Nicolos Vilant retires.

C.7 University of Aberdeen

Marischal College, Chair of Mathematics

1687 George Liddell appointed succeeding to his father Duncan.

1706 George Liddell deposed. Thomas Bower, Professor of Mathematics at King's College, Aberdeen, competes for the Chair, but Liddell is reinstated.

1715 Jacobite rebellion: the university is closed for two years.

1716 George Liddell deposed for his Jacobite sympathies.

1717 Colin Maclaurin appointed after competitive examination (Walter Bowman competes). Examiners: Charles Gregory, Professor of Mathematics at St Andrews and Alexander Burnet, regent at King's, Aberdeen.

1721–4 Colin Maclaurin travels in England and France.

1724 Daniel Gordon, a regent, teaches mathematics.

1725 Colin Maclaurin returns: in November he is appointed conjunt Professor with James II Gregory in Edinburgh.

* See the Gregory's family tree, p. 150.

1726 Colin Maclaurin deposed.

1727 John Stewart appointed. Examined by Daniel Gordon, Charles Gregory and Alexander Burnet (see above).

1753 Regenting abolished.

1766 Death of John Stewart.
William Trail(l) appointed after competitive examination (Robert Hamilton and John Playfair compete).

1768 John Gray, rector, leaves £1000 for two mathematical bursaries.

1775 Patrick Copland appointed to the Chair of Natural Philosophy.

1776 John Garioch appointed Trail's assistant and successor. He dies six months later.

1778/9 Patrick Copland teaches the higher branches of mathematics.

1779 William Trail resigns.
Patrick Copland appointed (but teaches natural philosophy).
Robert Hamilton appointed to the Chair of Natural Philosophy (but teaches mathematics).

1814 John Cruickshank teaches mathematics to the first and second class.

1817 Patrick Copland and Robert Hamilton exchange Chairs. John Cruickshank appointed Hamilton's assistant and successor. Hamilton continues teaching to the higher class up to 1824.

King's College, Chair of Mathematics

At King's during the eighteenth century, mathematics was taught by the regents. A Chair of Mathematics founded in 1703 was occupied by Thomas Bower, who neglected his duties, and by Alexander Rait, who was appointed to the Chair 'in order to gain him more respect from the students...but without any salarie' (quoted in Ponting (1979b), p. 172).

1703 Thomas Bower appointed Professor of Mathematics. His teaching discontinued.

1715 Jacobite rebellion: the university is closed for two years.

1717 Thomas Bower deposed.

1717–32 Chair vacant.

1732 Alexander Rait appointed Professor of Mathematics.

1753 Reform of the curriculum, but regenting is not abolished.

1800 Reform of regenting system. William Jack is appointed Professor of Mathematics.

1811 William Jack resigns; he becomes Professor of Moral Philosophy.

C.8 University of Dublin

Chairs of Mathematics and Astronomy

1762 Chair of Mathematics instituted.

1764 Richard Murray appointed Professor of Mathematics.

1774 Andrews Chair of Astronomy instituted.

1783 Henry Ussher appointed Andrews Professor of Astronomy.

1790 Death of Ussher.

      John Brinkley appointed Andrews Professor of Astronomy.

1791 Observatory at Dunsink completed.

1792 John Brinkley appointed Astronomer Royal of Ireland.

1795 Richard Murray retires from teaching.

1796 Bishop John Law institutes a prize to encourage the study of mathematics.

1799 Death of Murray.

1800 William Magee appointed Professor of Mathematics.

1812 Magee resigns.

1813 Bartholomew Lloyd appointed Professor of Mathematics.

APPENDIX D

MILITARY ACADEMIES

D.1 Royal Military Academy at Woolwich (masters and assistants of mathematics and fortification)

1741 John Muller, Chief Master (Benjamin Robins competes).
Derham, assistant to Chief Master.

1743 Death of Derham.
Thomas Simpson, assistant to Chief Master.

1761 Death of Simpson.
John Lodge Cowley, Assistant to Chief Master.

1764 William Green, master for classics, writings and arithmetic.

1766 Muller retires.
Allan Pollock, Professor of Artillery and Fortification.

1773 Cowley retires.
Charles Hutton, Professor of Mathematics (examined by Landen, Maskelyne, Horsley).

1777 Pollock retires.
Isaac Landmann, Professor of Artillery and Fortification.

1782 John Bonnycastle, mathematical master.

1789 Rouviere, assistant to Professor of Artillery and Fortification.

1792 Death of Rouviere.
C. Blumenheben, assistant to Professor of Artillery and Fortification.

1799 Green retires.
Louis Evans, mathematical master.

1802 Thomas Evans, first mathematical assistant.

1803 Olynthus Gregory, second mathematical assistant.

1806 Charles Martemont de Malorti, assistant to Professor of Artillery and Fortification.
Samuel Christie, third mathematical assistant.
Thomas Myers, fourth mathematical assistant.

Peter Barlow, fifth mathematical assistant.

William Moore, sixth mathematical assistant.

1807 Hutton retires.

John Bonnycastle, Professor of Mathematics.

William Saint, seventh mathematical assistant.

1816 Landmann retires.

Charles Martemont de Malorti, Professor of Artillery and Fortification.

Blumenheben retires.

David Robinson, assistant to the Professor of Artillery and Fortification.

1817 John Ritso, second assistant to Professor of Artillery and Fortification.

1821 Death of Bonnycastle.

Olynthus Gregory, Professor of Mathematics.

1838 Gregory retires.

D.2 Royal Military College (Sandhurst) (masters and professors of mathematics)

1799 Formation of the Senior Department.

Isaac Dalby appointed mathematical master.

1800 The staff consists of seven professors.

1802 Formation of the Junior Department. Sixteen cadets enroll.

Thomas Leybourn appointed mathematical master.

1803 Royal Warrant to extend the number of cadets to 400.

William Wallace appointed mathematical master.

1804 James Ivory appointed mathematical master.

1819 Ivory and Wallace resign (Wallace moves to the Chair of Mathematics in Edinburgh).

1839 Leybourn retires as senior Professor of Mathematics.

Other mathematicians belonging to the staff

Henry Clarke (1743–1818) retired in 1815.

James Cunliffe ⎫
Mark Noble   ⎬ contributors to Leybourn (1806–35).
John Wallace ⎭

John Lowry (1796–1847) retired in 1840.

D.3 Royal Naval Academy at Portsmouth (head-masters)

1733 Foundation of the Academy.

1735? Thomas Haselden appointed.

1740? Death of Thomas Haselden.
Robert Waddington appointed.

1755? Robert Waddington retires.

1755 John Robertson, former mathematical master at Christ's Hospital, London, appointed.

1766 John Robertson retires; he becomes clerk and librarian of the Royal Society.

1767 George Witchell appointed.

1771 John Bradley teaches mathematics as second mathematical master or 'mathematical usher'.

1785 Death of George Witchell.
William Bayley appointed.

1806 William Bayley retires.

1806/7 A reform takes place: the Academy becomes the Royal Naval College. James Inman appointed.

1829 Commissioned officers admitted in the College on half-pay.

1837 The College closed. James Inman retires.

APPENDIX E

SUBJECT INDEX OF PRIMARY LITERATURE

Algebra

Hales (1784)
Harris, John (1702)
Maclaurin (1784b)
Moivre (1708), (1724a)
Playfair (1779)
Saunderson (1740), (1756b)
Sewell (1796)
Taylor, Brook (1720a)
Trail (1796)
Waring (1762), (1764), (1766), (1770), (1772), (1779b)

Applications (engineering, gunnery, ship-building, fortifications, sound, geography, fluids, bridges, strength of materials, magnetism, navigation, geography)

Atwood (1794), (1796), (1798), (1801)
Barlow (1817), (1820)
Blake (1753), (1760)
Chapman (1820)
Euler (1776)
Glenie (1776)
Hales (1778)

Hawney (1717)
Hutton (1772), (1779a), (1779b), (1780), (1790), (1821)
Jurin (1744a), (1744b)
Muller (1746), (1747), (1755), (1757)
Perks (1717)
Playfair (1788b)
Robertson, John (1754)
Robins (1742), (1744)
Salmon (1749)
Young (1784)

Astronomy

Brinkley (1802a), (1803a), (1813), (1820a), (1820c), (1825)
Dawson (1769)
Emerson (1769b)
Gregory, David (1702)
Gregory, Olynthus (1802)
Hales (1782)
Harris, Joseph (1731)
Hellins (1798b), (1800)
Horsley (1768), (1770)
Ivory (1805), (1814), (1823)
Keill (1714), (1718)
Landen (1771b)
Laplace (1814)

Lax (1799)
Machin (1726), (1729), (1741)
Maclaurin (1754a), (1754b)
Milner (1780)
Murdoch (1753), (1769)
Newton, Isaac (1687), (1702)
Pemberton (1772)
Robertson, Abram (1807)
Silvabelle (1755)
Simpson (1756b), (1757), (1758)
Smeaton (1769)
Stewart (1756), (1761), (1763b)
Vince (1787), (1790), (1797,
1799, 1808)
Walmesley (1754), (1757),
(1759), (1762)
Whiston (1707)
Woodhouse (1812), (1818),
(1821)

Calculus

Barrow (1735)
Craig(e) (1706)
Ditton (1704)
Maclaurin (1742), (1744b),
(1744c)
Moivre (1698a), (1730)
Newton, Isaac (1687), (1704c),
(1711b), (1717), (1736), (1737),
(1745)
Pemberton (1724)
Simpson (1740)
Stirling (1717)
Taylor, Brook (1715), (1720b),
(1720c)
Waring (1776)
Woodhouse (1801b)

Elementary mathematics

Dalby (1806)
Fenn (1769, 1772)
Gregory, David (1745)
Hamilton, Robert (1800)
Hodgson (1723)
Holliday (1745)
Hutton (1798, 1801)
Inman (1810, 1812)
Martin (1736), (1739)
Muller (1748)
Simpson (1752)
Ward (1707)
West (1761)

Finite differences

Emerson (1763), (1767)
Gompertz (1806)
Moivre (1730)
Newton, Isaac (1687), (1711c)
Stirling (1720), (1730)
Taylor, Brook (1715)

Foundations of calculus

anonymous (1803a)
Bayes (1736)
Berkeley (1734), (1735a),
(1735b)
Blake (1741)
Carnot (1800, 1801)
Cheyne (1715)
Collins (1713)
Glenie (1778), (1789), (1793),
(1798)
Hales (1800)
Hanna (1736)
Heath (1752)
Horsley (1779a), (1779b)

Jurin (1734), (1735)
Kirkby (1748)
Landen (1758), (1764)
Ludlam (1770)
Maclaurin (1742)
Moivre (1704b)
Newton, Isaac (1687), (1704c), (1715)
Newton, Thomas (1805)
Paman (1745)
Petvin (1750)
Playfair (1822b)
Raphson (1715)
Robartes (1712)
Robins (1735)
Simson (1776)
Smith, James (1737)
Toplis (1805)
Walton (1735a), (1735b)
Woodhouse (1799), (1800)

Geometry

Ditton (1704)
Euclid (1756)
Hamilton, Hugh (1758)
Maclaurin (1720a), (1720b), (1720c)
Murdoch (1746)
Newton, Isaac (1687), (1704b)
Playfair (1795)
Simpson (1760)
Simpson (1724), (1776)
Stewart (1746), (1754), (1763a)
Stirling (1717)
Stone (1744)
Wallace (1798)
Waring (1765)

Integration

anonymous (1724)
anonymous (1803b)
Brinkley (1802b), (1803b), (1820b)
Bromhead (1816)
Cheyne (1703), (1705)
Cotes (1717), (1722)
Craig(e) (1685), (1688a), (1688b), (1693), (1695), (1698a), (1698b), (1699), (1704), (1710), (1718)
Hellins (1802), (1811)
Ivory (1798)
Klingenstierna (1733)
Knight (1812b)
Landen (1769), (1771a), (1772), (1775), (1780, 1789)
Landerbeck (1784)
Maclaurin (1742)
Moivre (1704a), (1717)
Newton, Isaac (1704c), (1736), (1745)
Perks (1708)
Robins (1728)
Simpson (1750b)
Vince (1786)
Wallace (1805)
Walmesley (1749b)
Woodhouse (1804)

Isoperimetrical problems, extremals, calculus of variations

Craig(e) (1702)
Fatio (1699), (1714)
Gregory, David (1698a), (1698b), (1700)
Machin (1720)
Newton, Isaac (1698)

Playfair (1812)
Sault (1699)
Simpson (1756a), (1759a)
Woodhouse (1810)

Mathematical dictionaries, collections

Bailey *et al.* (1736)
Barlow (1814a)
Clark *et al.* (1764, 1765, 1766)
Dodson (1748, 1753, 1755)
Harris, John (1704, 1710)
Hellins (1788)
Holliday (1745–53)
Hutton (1775a), (1775b), (1796, 1795)
Jones (1706)
Leybourn (1795–1804), (1806–35), (1817)
Maseres (1791–1807)
Raphson (1702)
Stone (1726)
Walter (1762?)

Mathematical tables

Barlow (1814b)
Inman (1829)
Knight (1817a)

Mechanics

Atwood (1784)
Ditton (1705)
Emerson (1754), (1769a)
Gregory, Olynthus (1806)
Keill (1710), (1717b)
Knight (1812a)
Landen (1777), (1780, 1789), (1785)

Maclaurin (1724)
Newton, Isaac (1687)
Parkinson (1785)
Robinson (1820)
Simpson (1750a), (1757)
Taylor, Brook (1714a), (1714b), (1715), (1723)
Venturoli (1822), (1823)
Vince (1781)
Waring (1788), (1789b)
Whewell (1819), (1823)
Wildbore (1791)
Wood (1796)

Methods of operators, Lagrangian methods

Babbage (1815), (1816), (1817a), (1817b), (1822), (1823)
Babbage and Herschel (1813)
Bonnycastle (1810), (1813)
Brinkley (1807)
Bromhead (1816)
De Morgan (1836)
Herschel (1814), (1816), (1818), (1822)
Knight (1811), (1816), (1817b), (1817c)
L'Huilier (1796)
Spence (1809), (1814), (1820)
Woodhouse (1801a), (1802), (1803)

Natural philosophy

Atwood (1776)
Cotes (1738)
Ditton (1705)
Hamilton, Hugh (1774)
Haüy (1807)
Helsham (1739)

Keill (1702), (1717a), (1720)
Maclaurin (1748a)
Newton, Isaac (1687)
Pemberton (1728)
Playfair (1812–14)
Robison (1804)
Rowning (1744, 1745)
Rutherforth (1748)
Whiston (1710)
Worster (1722)
Young (1800)

Optics

Smith, Robert (1738)
Stack (1793)

Perspective

Cowley (1765)
Kirby (1754)

Probability

Moivre (1712), (1718)
Simpson (1756b)

Series

Babbage (1819)
Bayes (1764)
Brinkley (1800c)
Craig(e) (1712)
Emerson (1767)
Hellins (1781), (1794), (1796),
(1798a), (1798b), (1800)
Hutton (1777), (1781)
Jones (1772)
Landen (1755b), (1761), (1780,
1789), (1781), (1783), (1784)
Maclaurin (1742)

Maseres (1777), (1779)
Moivre (1698b), (1699), (1724a),
(1730)
Montmort (1720)
Newton, Isaac (1711b)
Raphson (1697)
Robertson, Abram (1795), (1806)
Simpson (1740), (1743), (1753),
(1759b)
Stirling (1730)
Vince (1783), (1785)
Waring (1779a), (1784), (1786),
(1787), (1789a), (1791)

Shape of the Earth

Clairaut (1741a), (1741b), (1754)
Desaguliers (1726a), (1726b)
Ivory (1809), (1812a), (1812b),
(1822)
Maclaurin (1741), (1742),
(1744a)
Maupertuis (1733)
Newton, Isaac (1687)
Playfair (1805)
Short (1754)
Simpson (1743)
Stirling (1738)

Textbooks on calculus

Agnesi (1801)
anonymous (1810)
Babbage, Herschel and Peacock
(1820)
Cheyne (1703)
Colden (1751)
Dealtry (1810)
Ditton (1706)
Emerson (1743)
Harris, John (1702)

Hayes (1704)
Hodgson (1736)
Holliday (1777)
Hutton (1798, 1801)
Jones (1706)
Lacroix (1816)
Lardner (1825)
Lyons (1758)
Maclaurin (1742)
Martin (1736), (1739), (1759), (1773)
Muller (1736)
Rowe (1751)
Rowning (1756)
Saunderson (1756a)
Simpson (1737), (1750c), (1752)
Stone (1730)

Vince (1795)
Wallace (1810), (1815)
Ward (1707)

Trigonometry

anonymous (1724)
Brinkley (1800a), (1800b)
Cotes (1717)
Herschel (1813)
Landen (1755a)
Lyons (1775)
Martin (1773)
Moivre (1724b), (1730), (1741)
Pemberton (1722)
Woodhouse (1809)

APPENDIX F

MANUSCRIPT SOURCES

Cambridge University Library

MS Add 6312: Nicholas Saunderson's lectures on tides, capillary tubes, barometer, vapours, rain, hail, snow, figure of the Earth, thunder and lightening, winds, circulation of sap, rainbow, heat and cold, sound, chronology, astronomy, the horizontal Moon (dated 1737).

MS Add 2977: Nicholas Saunderson's lectures on hydrostatics, sound, optics, mechanics, astronomy, the tides, chronology, the horizontal Moon, heat and cold.

MS Add 589: Nicholas Saunderson's lectures on hydrostatics, sound, optics, mechanics, astronomy, the tides, chronology (dated 1727-9).

MS Add 3444: first part of Nicholas Saunderson's *Method of Fluxions* (dated 1738).

Gonville and Caius College Library (Cambridge)

MS Add 723/749: John Micklebourgh's mathematical exercise book (1720s).

Watson Library (University College, London)

MSS Graves 3-5: James Bradley's sketch of Oxford lectures

MS Add 243: Nicholas Saunderson's lectures on hydrostatics, the tides, sound, optics, mechanics, astronomy, the rainbow, chronology (dated 1730).

MS Graves 23 (2): letters to John Nourse from authors (John Stewart, John Colson, John Robertson, John Landen, William Emerson, Hugh Hamilton, Francis Holliday, Robert Heath, Robert Thorp, Nevil Maskelyne).

Bodleian Library (Oxford)

MS Rigaud 3–4: Nicholas Saunderson's lectures.

MS Radcliffe Trust: Thomas Hornsby Oxford lectures.

Royal Society Library (London)

MS LBC.16,412,426,456: Robert Simson's letters.

British Library (London)

MS Add Eg.834: Nicholas Saunderson's lectures.

MS Royal 487.b.17: first part of Nicholas Saunderson's *Method of Fluxions*.

NOTES

Introduction

1 This label for eighteenth-century calculus is in Boyer (1959), p. 224.
2 Charles Babbage in his (1864) recollects that he and his young fellows of the Analytical Society of Cambridge had in mind to entitle the first volume of their Memoirs *The Principles of Pure d-ism Against the Dot-Age of the University*. The volume appeared in 1813 with the more austere title *Memoirs of the Analytical Society*. This pun on the word 'dot' is widely known in the narrow circle of historians of mathematics.
3 On the introduction of Lagrangian ideas in early nineteenth-century England, see Koppelman (1971) and Enros (1981).
4 This was the title of a popular treatise by Benjamin Martin.
5 For example, Maclaurin (1742) was translated into French in 1749 and (partly) in 1765. Stone (1730) is a translation of L'Hospital's *Analyse des Infiniment Petits* (1696), and in 1801 there appeared an English translation of Agnesi's *Istituzioni Analitiche*.

Overture

1 See Newton (1967–81), II, pp. 206–47. The first edition of 'De analysi' is in Newton (1711a).
2 See Newton (1967–81), III, pp. 32–328. Newton (1736) is an English translation prepared by John Colson.
3 For an analysis of Newton's techniques of integration see Scriba (1964), pp. 125–6. See also Di Sieno and Galuzzi (1987).
4 See Newton (1967–81), III, pp. 102–4.
5 On the concepts of the differential calculus see Bos (1974).
6 According to Kitcher (1973) moments, fluxions and limits play different roles in Newton's mathematics. The fluxional method allowed him to reduce a diversified class of problems to two basic problems: finding tangents and finding areas. Moments were used to abbreviate calculations, while the rigorous proofs were framed in terms of the theory of limits. Kitcher concludes:

Once we have unearthed the problems, Newton's solutions seem more impressive than his eighteenth-century critics took them to be. (Kitcher (1973), p. 49)

Chapter 1

1 I summarize here the main events of this famous quarrel. As is well known, in 1676 Oldenburg promoted an exchange of letters between Newton and Leibniz. Newton disclosed part of his secrets, especially his progress on series, in two letters, known later as *epistola prior* and *epistola posterior*, which were first published by John Wallis (1699)

in the *epistolarum collectio* included in the third volume of his mathematical works. By then Leibniz had already independently 'invented' the differential calculus. Leibniz's 'new method' appeared in (1684), while Newton's calculus was rendered public only in 'De quadratura' (1704c). Leibniz reviewed anonymously the 'De quadratura' in the *Acta Eruditorum* for January 1705. He subtly affirmed that Newton had only modified details of Leibniz's calculus. To be fair we must say that Leibniz had already been attacked by the Newtonian Fatio in (1699): so his 1705 review can be seen as a self defence. But the situation precipitated when another Newtonian, John Keill, in (1710) bluntly accused Leibniz of plagiarism. Leibniz's next move was to ask for a public apology from the Royal Society. A committee, completely under Newton's influence, was immediately formed and, early in 1713, appeared the result: the *Commercium Episolicum*, a collection of letters and mathematical writings which 'proved' Leibniz's plagiarism. Leibniz's reply was to circulate a *charta volans*, a 'fly-sheet', which included a *Judicium* by an 'eminent mathematician'. Johann Bernoulli, who maintained that Newton could not understand second order differentials. Newton counter-attacked with the 'Account' to the *Commercium Episolicum*, published in 1715. Leibniz's death in 1716 did not calm Newton's anger: in 1717 and 1722 there appeared Newton's revised and augmented editions of Raphson's *History of Fluxions* (1715) (in which Newton added two letters by Leibniz to Conti of 1716, his response to the former and comments upon the latter) and of the Royal Society's *Commercium Episolicum* (in which Newton added the 'Ad Lectorem', a latin translation of his 'Account' (1715), the offending 1713 *Judicium* together with Newton's annotations). It is now accepted that Newton is the 'first inventor', but it is certain that Leibniz worked independently on his differential and integral calculus.

2 On David Gregory see Eagles (1977a) and (1977b).

3 These pages were communicated in 1692 to Wallis by Newton. They are reproduced in Newton (1967–81), VII, pp. 170–80.

4 Gregory's copy of Newton's 'De methodis' is in Newton (1967–81), III, pp. 354–72.

5 Some of these papers relate to the challenges between the British and the continentals. As is well known the Newton–Leibniz controversy provoked a great rivalry between the two mathematical communities. A first challenge was proposed by Johann Bernoulli. In the summer of 1696 he proposed to Leibniz, Varignon, L'Hospital, and others, the famous 'brachistocrone' problem: i.e. to find, given two points A and B in a gravitational field, the path which minimizes the time of fall of a point mass. In September 1696 this problem reached Wallis in Oxford. Wallis after three months passed it to his colleague David Gregory, who could not solve it. At last, in January of the next year, Bernoulli's challenge reached Newton who could, in a few hours, understand that the brachistocrone is a cycloid. Newton's solution appeared anonymously as [Newton] (1698). In Fatio (1699) there was an original approach to the problem: Fatio's solution was later perfected by Newton. Craig(e) (1702) presented 'his own' analytical solution in differential notation. Machin (1720) tried unsuccessfully to extend the study of the brachistocrone in a central gravitational field: his mistakes are analysed in Newton (1967–81), VIII, p. 13n. Craig(e) kept on using the differential notation: only in (1712) and (1718) he employed Newton's dots. David Gregory (1698a) gave another solution of the brachistocrone, probably achieved only with Newton's assistance. Another paper on the brachistocrone is Sault (1699). Another challenge was raised in 1716, when the controversy between Newtonians and Leibnizians was at its climax. It was poorly treated in the anonymous [Newton] (1717). The challenge consisted in the following problem: to find the orthogonal trajectories to a given family of curves defined as $f(x,y,a) = 0$, where a is a parameter, and x and y are the coordinates which vary on a curve of the family when a is constant. In January 1716 Leibniz posed the problem to the British mathematicians in a letter to Conti. The general problem was exemplified in Leibniz's letter by a family of hyperbolas of varying *latus rectum*. The Newtonians John Keill and the young James Stirling in Oxford, and de Moivre, John Machin, Henry Pemberton, Brook Taylor in London did not understand the generality of the problem and found the orthogonals to the family of hyperbolas. Keill's solution is in Newton (1959–77), VI, pp. 282–3; Machin's is in Rigaud

(1841), I, pp. 268–9; Stirling's is in the Appendix of his book (1717); Pemberton's has been found in manuscript by D. T. Whiteside in the University Library of Cambridge (see Newton (1967–81), VIII, p. 63n); while, as far as we know, de Moivre's and Taylor's solutions have not survived. In the early Spring of 1716 Leibniz posed in a letter to Conti a new problem on the orthogonal trajectories to an arbitrary family of curves defined by a parametric differential condition. The only British solution was Taylor (1720b), elegant but derived from Hermann. It is interesting to note that the Newtonians failed to answer the 1716 challenge: a problem which involved the use of partial derivatives. In fact in this context one has to deal both with $\partial f/\partial x$ and with $\partial f/\partial a$. As we will see later, the Newtonians failed to understand the importance of the calculus of partial derivatives developed by the continentals. On this topic see Engelsman (1984). For further information on the challenges between the Newtonians and the Leibnizians see Whiteside's notes in Newton (1967–81), VIII, and Hall (1980).

6 'Niewentiit' is the Dutch theologian Bernhardt Nieuwentijdt, who wrote several works in criticism of the differential calculus. L. Carré is the author of *Méthode pour la Mesure des Surfaces, la Dimension des Solides...par l'Application du Calcul Intégral* (1700).

7 For example, we are told in Trail (1812), p. 2n, that Robert Simson studied Jones (1706) in Glasgow.

8 The 'iatro-mechanists' tried to extend Newton's mathematical laws to the study of medicine.

9 Cheyne replied in (1705) and then, with the exception of a philosophical treatment of the infinite in his (1715), abandoned mathematics. On the Cheyne affair see Whiteside's reconstruction in Newton (1967–81), VIII, pp. 15–21.

10 It is interesting to note that Hayes, as Harris (1702), 2nd edn, refers the reader to continental mathematicians. In the Preface Hayes lists Wallis, Barrow, Newton, Leibniz, L'Hospital, 'The Bernoullis', Craige, Cheyne, James Gregory, Tschirnaus, de Moivre, Fatio de Duillier, Varignon, Nieuwentijdt, Carré.

11 The Mathematical School at Christ's Hospital was established in 1673. The discipline in the School was notoriously a disaster. An attempt of reform took place in 1694: John Wallis, David Gregory and Isaac Newton were consulted. However, the School never fulfilled the desiderata of the founders, most notably amongst whom was Samuel Pepys. The problems at Christ's Hospital were caused by the Masters, Edward Paget and Samuel Newton, rather than by the pupils. In 1709 James Hodgson (1672–1755) was appointed; we will encounter him in chapter 4 as a writer on fluxions. I believe that he retired from teaching in 1748. His successor, John Robertson (1712–76), was a mathematical writer: he became in 1755 first master at the Royal Naval Academy at Portsmouth: see his *Elements of Navigation* (1754). In 1755 he was succeeded by James Dodson (1709–57), Augustus De Morgan's ancestor: among his works we should at least remember *The Mathematical Repository* (Dodson, 1748, 1753, 1755), almost a bible for the 'philomaths'. After Dodson we can identify two other masters: Daniel Harris, who was at Christ from the 1750s to the 1770s, and William Wales (1734–98), who was appointed in 1775. The *New* Mathematical School, where Humphry Ditton was master from 1706 to 1715, lasted only his lifetime.

12 Stone was a prolific scientific writer. In addition to the works already mentioned, he revised the second edition of the English translation of David Gregory (1702). His translation of Barrow's *Lectiones Geometricae* appeared as Barrow (1735). Among his other works we mention his translation of Nicolas Bion's treatise on mathematical instruments and L'Hospital's treatise on conic sections.

13 In the section devoted to 'Astronomy' Hayes mixed propositions taken from Newton's *Principia* with propositions taken from Leibniz's *Tentamen de Motuum Coelestium Causis*. See Hayes (1704), pp. 291–305.

14 In October 1700 Dr Charlett, Master of University College, Oxford, sent to Pepys a 'Scheme' of Gregory's course of lectures on the 'Elements of Mathematical Science' intended for classes of ten or fifteen 'scholars'. It seems that in the universities, both of Scotland and England, courses were given not only to the young students, but also to

adults interested in science. Gregory's 'Scheme', which can be found in Howson (1982), pp. 42–3, included only Euclid's *Elements*, trigonometry and algebra as a mathematical background for a discussion on optics, astronomy and mechanics.

15 MSS Graves 3–5 (University College Library, London) provide just an outline of the lectures which Bradley had in mind to give. It is interesting however to see the texts which Bradley was using. Thomas Hornsby continued in the second half of the century to lecture in the Ashmolean Museum. He left about 2500 manuscript pages of lectures which are kept in the Bodleian Library (Oxford) as MS Radcliffe Trust. The historian of mathematics will find very little in these manuscripts.

16 The Savilian Chair of Geometry after Edmond Halley passed to Nathaniel Bliss (1700?–64) who taught some elementary mathematics (see Turner (1986), pp. 677–8). He was succeeded by Joseph Betts and John Smith, the latter being an anatomist and chemist rather than a mathematician. Abram Robertson, the next to occupy the Chair of Geometry, wrote a treatise on conic sections and some papers on the binomial theorem (1795, 1806) and astronomy (1807).

17 On Whiston see Force (1985).

18 Waterland's and Greene's plans of courses are reproduced in Wordsworth (1877).

19 Smith patronized the study of mathematics in various ways. In his will he left to the University of Cambridge £3500, part of which served to support the Smith prize (see 'Smith Robert' in *Dictionary of National Biography*).

20 Cambridge University Library (Cambridge) Add MSS 6312, 2977, 589; Bodleian Library (Oxford) MS Rigaud 3–4; University College (London) MS Add 243; British Library (London) MS Add Eg.834.

21 The manuscript of the first part is in Cambridge University Library (Cambridge) MS Add 3444 and British Library (London) MS Royal 487.b.17.

22 See chapter 2 and, for further details, Gowing (1983).

23 For a rare record of the teaching of fluxions in Cambridge in the 1720s see 'John Micklebourgh's Mathematical Exercise-Book', Gonville and Caius College Library (Cambridge), MS Add 723/749.

24 John Robison, who had been a pupil of Simson, states that Simson gave lectures on 'the elements of the fluxionary calculus' (*Encyclopaedia Britannica*, 3rd edn., XVII, pp. 504–9).

25 Royal Society MS LBC.16,412,426,456.

26 It seems that Maclaurin took his teaching duties in Edinburgh very seriously. In his correspondence he describes himself as being occupied six hours a day teaching to more than one-hundred pupils. See Maclaurin (1982), pp. 26, 32, 148.

27 *The Scots Magazine. Containing a General View of the Religion, Politicks, Entertainment, &c. in Great Britain: And a succint Account of Publick Affairs Foreign and Domestick. For the Year MDCCXLI*, III, by Sands, Brymer, Murray, Cochran, 1741, p. 372.

28 St Andrews and Aberdeen do not compare with Glasgow and Edinburgh in this early period. In St Andrews the Gregorys dominated the Chair of Mathematics from 1688 to 1763. But we do not know very much about the teaching of Charles and David II Gregory. In King's College, Aberdeen, a Chair of Mathematics was established at the beginning of the century, but it was not occupied. In Marischal College, Aberdeen, Colin Maclaurin held the Chair of Mathematics from 1717 to 1725, but he did not teach. In 1727 he was succeeded by John Stewart who translated Newton (1704c) and (1711b) into English as Newton (1745).

Chapter 2

1 On Cotes see Gowing (1983). This work contains a detailed commentary and a translation of Cotes's 'Logometria' (1717). See also anonymous (1724), a review, possibly by Robert Smith, of Cotes's *Harmonia Mensurarum* (1722).

2 Cotes's *Hydrostatical and Pneumatical Lectures* (1738) were published posthumously and became a standard textbook, still in use in Cambridge in the second half of the eighteenth century.

3 All these results have been analysed in detail in Gowing (1983).

4 These eighteen 'forms' are reproduced in Gowing (1983), pp. 192–4.
5 Cotes's factorization theorem was enunciated without proof by Robert Smith as follows:

Si quaerantur Factores Binomii $a^\lambda \pm x^\lambda$. Indice λ existente quolibet integro: dividatur circuli circumferentia $ABCD$ cujus centrum O, in totidem partes aequales AB, BC, CD, DE, EF, &c; quot sint unitates in 2λ; & ab omnibus divisionibus ad punctum quodvis P in OA radio si opus producto situm, ducantur rectae AP, BP, CP, DP, EP, FP, &c; deinde positis $OA = a\ OP = x$, contentum sub omnibus AP, CP, EP, &c. sumptis a divisionibus alternis per integrum circuitum, adaequabit $a^\lambda - x^\lambda$ vel $x^\lambda - a^\lambda$, prout P fuerit intra vel extra circulum: & contentum sub reliquis BP, DP, FP, &c. in locis reliqui alternis adaequabit $a^\lambda + x^\lambda$. (Cotes (1722), pp. 113–14)

A proof was given in Pemberton (1722), Moivre (1730) and Brinkley (1800b).
6 This is form XLVI. See Cotes (1722), p. 187. Smith is also the author of the 'Editoris Notae', placed at the end of *Harmonia Mensurarum* (1722), pp. 93–125, in which he employs Cotes's integrals to solve some problems of the 'Logometria' (1717).
7 For a detailed analysis of de Moivre's life and work see Schneider (1968).
8 Taylor (1715) has been analysed in detail in Feigenbaum (1985).
9 Another example of the flexibility of Newton's notation is given in Brinkley (1807) where some theorems of the Lagrangian calculus of operations (quite obviously an approach of which Taylor was not aware) are expressed in dotted notation. See chapter 9, section 9.4.
10 The original Latin reads as

Sint z & x quantitates duae variabiles, quarum z uniformiter augetur per data incrementa $\dot{z}$, & sit $n\dot{z} = v$, $v - \dot{z} = \dot{v}$, $\dot{v} - \dot{z} = \ddot{v}$. & sic porro. Tum dico quod quo tempore z crescendo fit $z + v$, x item crescendo fiet

$$x + x\frac{v}{1 \cdot \dot{z}} + x\frac{\dot{v}v}{1 \cdot 2 \cdot \ddot{z}^2} + x\frac{\dddot{v}vv}{1 \cdot 2 \cdot 3 \cdot \dddot{z}^3} + \dots.$$

11 The original Latin reads as

Si pro Incrementis evanescentibus scribantur fluxiones ipsis proportionales, factis jam omnibus $\dot{v}$, $\dot{v}$, v, v, v, &c. aequalibus quo tempore z uniformiter fluendo fit $z + v$ fiet x,

$$x + \dot{x}\frac{v}{1 \cdot \dot{z}} + \ddot{x}\frac{v^2}{1 \cdot 2 \cdot \dot{z}^2} + \dddot{x}\frac{v^3}{1 \cdot 2 \cdot 3 \cdot \dot{z}^3} + \dots.$$

12 Tweedie does not accept this interpretation; see Tweedie (1922), p. 8. According to Peter J. Wallis, in 1717 Stirling might have held a teaching appointment in Edinburgh: see 'Stirling, James' in *Dictionary of Scientific Biography*.
13 A letter by Stirling to Newton is dated Venice 17 August 1719. See Newton (1959–77), VII, pp. 53–4.
14 For Watts' Academy (known up to 1721 as 'Accomptant's Office') see Hans (1951), pp. 82–7. Stirling, together with W. Watts, W. Vream and P. Brown, wrote a textbook on mechanics and natural philosophy to be used in the Academy. Another textbook is Worster (1722). A study of Stirling's life could tell us something about the quasi-professionalization of mathematics in the first half of the eighteenth century. It seems that in Venice he was involved in an obscure affair in the glass industry; then we find him in London as a teacher of mathematics and adviser to Lord Bolingbroke on financial calculations, and in Scotland as an administrator of the lead mines, a surveyor and, perhaps, a teacher of book-keeping, navigation, practical mathematics, georgraphy and French. Furthermore, Stirling's biography is remarkable because it shows that amongst British mathematicians he was the one who had most contacts with the continent. On Stirling see Tweedie (1922). This work includes a biography of Stirling, a commentary on Stirling (1717) and Stirling (1730) and part of his correspondence with Colin Maclaurin (1698–1746), Gabriel Cramer (1704–52), Nicolas Bernoulli (1687–1759), Luis-Bertrand Castel (1688–1757), James Bradley (1693–1762), Samuel Klingenstierna (1698–1765), John Machin (d.1751), Alexis-Claude Clairaut (1713–65), Leonhard Euler (1707–83) and Martin Folkes (1690–1754).

15 The original Latin reads as

Quippe ubi z est quantitas parva, prior forma erit adhibenda; & posterior ubi magna. Et hae Series quae componuntur ex Factoribus in progressione Arithmetica, longe magis idoneae sunt huic negotio quam vulgares quae constantur ex dignitatibus indeterminatae ascendentibus vel descendentibus.

16 Stirling tabulates the Γ_n^s (Stirling's numbers of the second species) and the C_n^r (Stirling's numbers of the first species) on p. 8 and p. 11 of his (1730).

17 Some misprints have been corrected.

18 The original Latin reads as

Propono jam inventionem Termini qui consistit in medio inter duos primos 1 & 1. Et quoniam Logarithmi Terminorum initialium habent differentias lente convergentes, primum quaeram Terminum in medio consistentem inter duos ab initio satis remotis, verbi gratia, inter decimum primum 3628800 & decimum secundum 39916800: & ex eo dato regrediar ad Terminum quaesitum.

19 The original Latin reads as

Unde constat Terminum inter 1 & 1 esse .8862269251; cujus quadratum est .7853...&c, scilicet Area Circuli cujus Diameter est unitas. Atque illius Termini duplus 1.7724538502 [...] aequalis est Radici quadrati numeri 3.1415926...&c. qui denotat Circumferentia Circuli, cujus Diameter est unitas.

20 This is Stirling's proof:

Minuatur variabilis z decremento suo 2n; vel quod idem est, substituatur $z-2n$ pro z in Serie

$$\frac{z\log(z)}{2n} - \frac{az}{2n} - \frac{an}{12z} + \frac{7an^3}{360z^3} - \frac{31an^5}{1260z^5} + \dots$$

& provienet valor ejusdem successivus

$$\frac{(z-2n)\log(z-2n)}{2n} - \frac{a(z-2n)}{2n} - \frac{an}{12(z-2n)} + \frac{7an^3}{360(z-2n)^3} - \dots.$$

Hunc subducto de valore priore, Terminis prius ad eandem formam per Divisionem reductis, & reliquetur

$$\log(z) - \frac{an}{z} - \frac{an^2}{2z^2} - \frac{an^3}{3z^3} - \frac{an^4}{4z^4} - \dots$$

id est, Logarithmus numeri $z-n$. Adeoque universaliter decrementum duorum valorum successivorum Seriei, aequatur Logarithmo ipsius $z-n$; qui exprimit in genere quamvis Logarithmorum qui erant summandi (Stirling (1730), p. 136).

de Moivre expressed $\log(x!)$ as follows (see Tweedie (1922), p. 43):

$$\tfrac{1}{2}\log(2\pi) + (x+\tfrac{1}{2})\log x - x$$

$$+ \frac{B_1}{1\cdot 2}\frac{1}{x} - \frac{B_2}{3\cdot 4}\frac{1}{x^3} + \dots + (-1)^{n+1}\frac{Bn}{(2n-1)^{2n}}\frac{1}{x^{2n-1}}\dots,$$

B_1, B_2, etc., being the Bernoulli numbers (see Moivre (1730), pp. 99ff and (1730) 'Supplementum', pp. 1–18).

21 Stirling's analytic treatment of cubics was followed by Maclaurin (1720c) devoted to the organic description of curves, and by Patrick Murdoch (1746). These became three classic commentaries of Newton's 'Enumeratio' (1704b). In the last two the calculus of fluxions was not employed.

22 In a letter to Newton (29 December 1716) John Keill wrote that Stirling had solved the problem of orthogonal trajectories, see Newton (1959–77), VI, pp. 282–3. Indeed, like all British mathematicians, Stirling did not solve the general case.

23 On Maclaurin see Tweedie (1915), Turnbull (1951) and Maclaurin (1982).
24 This was published as Maclaurin (1724).
25 It is interesting that a use of infinitesimals is found in the early work of Maclaurin. He later became a great adversary of 'infinites'. Maclaurin's shift is typical of the development of the fluxional calculus in the first half of the century. See chapter 3, section 3.4.
26 This belief is reported, for instance, in D. Gregory (1702), *Prefatio*, where it is stated that the 'ancients' knew the inverse square law and the concept of universal gravitation.

Chapter 3

1 I will not deal here with Berkeley's explanation of the calculus in terms of compensation of errors. This idea was reconsidered by Lazare Carnot.
2 Only a few extremists, such as William Emerson (see section 3.3 below) and John Colson (see chapter 4, section 4.1) could think of mathematics as a strictly empirical science.
3 As far as I know the compensation of errors was not considered favourably by any British mathematician.
4 In his *Analysis Aequationum Universalis* (1697), Raphson developed several techniques of approximating the roots of algebraical equations and the so-called Newton–Raphson method, and he discussed the concept of mathematical infinites.
5 For instance, see in Stirling (1717) the free use of infinitely little and infinitely large quantities.
6 The original Latin reads as

Hinc ideam habemus analogiae quae est inter Methodum Differentialem & Methodum vulgarem Serierum; haec procedit per Fluxiones sive rationes differentiarum ultimas, illa vero generaliter per differentias cujuscumque magnitudinis.

7 There was a dispute between the fluxionists, each one claiming to have the right answer to Berkeley. It is not my purpose to enter into the details of this quarrel which, strangely enough, has been given great importance in Cajori (1919). The texts are the following: Berkeley (1734), (1735a) and (1735b); Jurin (1734) and (1735); Walton (1735a), (1735b); Robins (1735); Bayes (1736); Hanna (1736); and J. Smith (1737). The polemic was continued in a periodical called *The Present State of the Republick of Letters*, for W. Innys and R. Manby, London. The papers on the calculus written by Jurin, Robins and Pemberton are from October 1735 to December 1736. This journal was continued as *The Works of the Learned*, in which, from February 1737 to October 1737, Jurin and Pemberton continued the debate. Paman (1745), Petvin (1750) and Heath (1752) can be read in connection with this dispute. The list of 'answers' to Berkeley is endless. Even seventy years later the fluxionists continued this sterile exercise: e.g. Hales (1800) and T. Newton (1805).
8 For instance, Jurin wrote:

Where, said I, do you find Sir Isaac Newton using such expressions as *the velocities of the velocities, the second third and fourth velocities, the incipient celerity of an incipient celerity, the nascent augment of a nascent augment?* Is this the true and genuine meaning of the words fluxionum mutationes magis aut minus celeres? (Jurin (1735), p. 35).

9 I am quoting from *Sir Isaac Newton's Mathematical Principles of Natural Philosophy and his System of the World*, Motte's translation, revision and notes by Cajori, Berkeley and Los Angeles: University of California Press, 5th printing, 1962, p. 38.
10 It is worthwhile quoting Jurin's answer to Berkeley on this specific point:

I forbear making any remarks upon your interpretation of the word vanish. I admit it to be as you are pleased to make it, that the first supposition is, there are increments; and that the second supposition is, there are no increments. What do you infer from this? The second supposition, say you, is contrary to the former, and destroys the former, and in destroying the former it destroys the expressions, the propositions, and everything else derived from the former supposition. Not too fast Good Mr. *Logician*. If I say, the increments now exist, and the increments do not now exist; the latter assertion will be contrary to the former, supposing now to mean the same instant of time in both assertions. But if I say at one time, the increments now exist; and say an hour after, the increments do not now exist; the latter assertion will neither be contrary, nor contradictory to the former, because the first now signifies one time, and the second now signifies another time, so that both assertions may be true. The case therefore in your argument does not come up to your *Lemma*, unless you will say Sir Isaac Newton supposes that there are increments, and that there are no increments, at the same instant of time. Which is what you have not said, and what, I hope, you will not dare to say.

But perhaps you will still maintain, that whether the second supposition be esteemed contrary, or not contrary, to the first, yet as the increments, which were at first to exist, are now supposed not to exist, but to be vanished and gone, all the consequences of their supposed existence, as their expressions, proportions, &c. must now be supposed to be vanished and gone with them. I cannot allow of this neither.

Let us imagine yourself and me to be debating this matter, in an open field, at a distance from any shelter, and in the middle of a large company of Mathematicians and Logicians. A sudden violent rain falls. The consequence is, we are all wet to the skin. Before we can get to covert, it cleans up, and the Sun shines. You are for going on with the dispute. I desire to be excused, I must go home and shift my clothes and advise you to do the same. You endeavour to persuade me I am not wet. The shower, say you, is vanished and gone, and consequently coldness and wetness, and every thing derived from the existence of the shower, must have vanished with it. I tell you I feel my self cold and wet. I take my leave, and make haste home. I am persuaded the Mathematicians would all take the same course, and should think them but very indifferent Logicians, that were moved by your arguments to stay behind. (Jurin (1735), pp. 96-8).

11 I am quoting from *Sir Isaac Newton's Mathematical Principles of Natural Philosophy and his System of the World*, Motte's translation, revision and notes by Cajori, Berkeley and Los Angeles: University of California Press, 5th printing, 1962, pp. 38-9.

12 The project of making fluxions 'visibles, or even tangibles' and to proceed by 'ocular demonstrations' was advanced by John Colson in his explanatory notes to Newton (1736). See chapter 4, section 4.1.

13 James Jurin was educated at Christ's Hospital, Cambridge and Leyden. He was head-master of Newcastle Grammar School from 1710 to 1715. He was also Secretary of the Royal Society from 1721 to 1727. Benjamin Robins taught mathematics in London, and dedicated himself to questions concerning artillery and fortifications. In 1741 he was unsuccessful in his attempt to be elected first master at the Royal Military Academy at Woolwich. He died in India, where he was employed, from 1749, as engineer-general of the East India Company. For Robins's work on artillery see Robins (1742), reprinted with additional tracts in Robins (1761).

14 The confrontation between Robins and Jurin is in *The Present State of the Republick of Letters*, XVI (Jul. 1735-Dec. 1735); XVII (Jan. 1736-Dec. 1736).

15 On the limit-avoiding *versus* limit-achieving interpretation of limits, see Grattan-Guinness (1969).

16 Maclaurin (1742) is not just a work written in answer to Berkeley's criticisms. In 763 pages Maclaurin covers not only the foundations of the calculus, but also treats extensively the theory and the applications of the calculus of fluxions. Particularly valuable are: chapter IX, Book 1, on the application of the Taylor expansion in the study of maxima, minima and points of inflexion; chapter X, Book 1, on the Euler-Maclaurin summation formula; chapter XIII, Book 1, on isoperimetrical problems; chapter XIV, Book 1, where Maclaurin reproduced a corrected version of his prize essay on tides (1741); and chapter III, Book 2, which stimulated Landen's research (1775) on integrals. Some aspects of Maclaurin's mathematics are treated in Tweedie (1915), Turnbull (1951) and Scott (1971).

17 Berkeley had already detected the vicious circle:

It must, indeed, be acknowledged, that he [Newton] used Fluxions, like the Scaffold of a building, as things to be laid aside or got rid of, as soon as finite Lines were found proportional to them. But then these finite Exponents are found by the help of Fluxions. Whatever therefore is got by such Exponents and Proportions is to be ascribed to Fluxions: which must therefore be previously understood. (Berkeley (1734), pp. 58–9).

18 Forms very similar to these axioms had already been studied by Bayes (1736), p. 18.
19 'Hae Geneses in rerum natura locum vere habent'; Newton (1704c), p. 165.

Chapter 4

1 The following should also be noted: Martin (1736), (1759) and (1773), and Colden (1751), had a chapter on fluxions; Hales (1800) was a pompous Latin work mainly concerned with foundations; and Agnesi (1801) was a translation of the famous Italian treatise.
2 The editor of Newton (1737) is unknown. It might have been James Wilson (1690–1771). See Wallis and Wallis (1986), p. 109.
3 As appears from appendix A.1, in the second edition of Emerson (1743) some problems on physical astronomy are added.
4 Quoted in Clarke (1929) from the Simpson's papers, Columbia University Library, New York.
5 Quoted in Clarke (1929) from the Simpson's papers, Columbia University Library, New York.
6 That is: Hodgson (1736; 2nd edn. 1756; 3rd edn. 1758); Muller (1736); Newton (1736), (1737), (1745); Simpson (1737), (1750c; 2nd edn. 1776); Blake (1741; 2nd edn. 1763); Maclaurin (1742); Emerson (1743; 2nd edn. 1757; 3rd edn. 1768; 4th edn. 1773); Rowe (1751; 2nd edn. 1757; another? 1762; 3rd edn. 1767); Rowning (1756); Saunderson (1756a); Lyons (1758); Holliday (1777). From 1777 to 1795 there was a period of silence, broken only by the edition of Newton's *Opera* (1779–85) and the second edition in 1792 of Simpson (1752). Then in the last years of the century and the first decades of the nineteenth century we find an increase in publications. Vince (1795) ran to five editions and Dealtry (1810) ran to a second edition in 1816. In 1801 we find the second edition of Maclaurin (1742) and Agnesi (1801); in 1809 the fourth edition of Rowe (1751) and the third of Blake (1741); in 1805 and 1823 the third and fourth editions of Simpson (1750c).
7 For John Nourse Emerson wrote: *Cyclomathesis* (1763–91), a course consisting of many volumes to which belonged *The Arithmetic of Infinites* (1767) as volume V, *Mechanics; or the Doctrine of Motion* (1769a) as volume VII, and *A System of Astronomy* (1769b) as volume VIII. *Tracts* (1793) was published for F. Wingrave, successor to John Nourse. On John Nourse see Feather (1981).
8 On Simpson see Hutton (1792) and Clarke (1929).
9 Agnesi (1801) was revised by John Hellins.
10 Hodgson was a *protégé* of Flamsteed: he married Flamsteed's niece and was co-editor of Flamsteed's *Atlas Coelestis*.
11 Simpson (1752) was reissued in 1792 and 1810. It was also translated into French.
12 On Hutton's biography see Howson (1982), pp. 59–74. See also anonymous (1805),
.   Bruce (1823) and O. Gregory (1823).
13 On the Tripos Exam see Ball (1880) and Gascoigne (1984).
14 This letter quoted in Gascoigne (1984) is from *The Connoisseurs*, CVII, Thursday 12 February 1756. I am quoting from *The Connoisseurs. By Mr. Town, Critic and Censor-General*, 6th edn., vol. IV, *for* J. Rivington *et al.*, Oxford, published in 1774, pp. 18–24. The letter, probably written by the editor for the amusement of his readers, is signed 'B.A.'.
15 However, a distinction should be made between the geometric classicism of Robert Simson and Matthew Stewart (who were isolated from the learned society of common-sense philosophers), John Robison's criticisms of the analytic methods in *mechanics* and Dougald Stewart's perplexity toward analytic methods in *algebra*. Furthermore, a key figure in late eighteenth-century Scottish science such as John Playfair was, as we will see in chapter 7, section 7.2, very interested in introducing analytical techniques both at

a research and at a teaching level. The situation in Scotland was therefore more varied than is usually believed: together with 'pure geometricians' there were scientists quite prepared to accept the 'analysis' of the continentals.

16 On the 'philomaths' see Pedersen (1963), P. J. Wallis (1973) and Wallis and Wallis (1986).

17 The members of the famous Lunar Society of Birmingham devoted attention to experimental science rather than mathematics.

18 Maclaurin with his *Account of Sir Isaac Newton's Philosophical Discoveries* (1748a) is unusually deep and thoughtful: his last effort is a philosophically valuable analysis of the methodology of the Newtonian mathematization of mechanics and astronomy. However, generally the introductions to natural philosophy, even though longer in comparison with Whiston's and Keill's lectures, did not invite the reader to study any philosophical or mathematical problem. A famous case of how this kind of simple popularization of science could become extremely profitable is that of Benjamin Martin (1704–82), one of the most successful itinerant mathematics teachers and demonstrators of experiments. His vast production of popular scientific books was exceedingly derivative. In Martin (1736), (1739), (1759) and (1773) there are some sections on the fluxional calculus. On Martin see Millburn (1976) and (1986). Also, the increasing success of scientific dictionaries and encyclopaedias is a measure of the interest in a simple approach to science: beside the famous Ephraim Chamber's *Cyclopaedia*, there were the second edition of Edmund Stone's *Mathematical Dictionary* (1726) in 1743, Temple Henry Croker, Thomas Williams and Samuel Clark's *Complete Dictionary of Art and Sciences* (Clark *et al.* (1764, 1765, 1766)), which was derived from Chamber and the French *Encyclopédie*, and Thomas Walter's *A New Mathematical Dictionary* (1762?).

19 I have found in MS Graves 23 (2), University College Library (London), several letters to John Nourse from authors of scientific books. These should be read together with the manuscripts described in Feather (1981). Useful information on eighteenth-century publishers can be found in Rivers (1982); see especially Rousseau's contribution (1982).

Chapter 5

1 On the development of fluid mechanics see Truesdell (1954); on the shape of the Earth see Todhunter (1873) and Aiton (1955); and on the Moon's orbit see Waff (1976). In this chapter I will concentrate on the researches by Simpson and Maclaurin on the attraction of ellipsoids. The contributions of the British in the field of physical astronomy were negligible in comparison with contemporary continental research. I will give in this note some information on the researches of the fluxionists on planetary motions.

Charles Walmesley, Thomas Simpson and Matthew Stewart were the three British mathematicians who wrote on physical astronomy around the middle of the century.

Charles Walmesley ('Pastorini') (1722–97) was a Roman Catholic prelate who was educated in the English Benedictine College of St Gregory at Douai and at the Sorbonne. He was consecrated Bishop of Rama. His career in the Church culminated in 1770 when he became vicar-apostolic. He died in 1797 at Bath. His first works (1749a) and (1749b) were devoted to the theory of the Moon and to Cotes's integrals. Walmesley (1757) treated the precession of the equinoxes, following some theorems on the same subject by Silvabelle (1755). Both papers were criticized in a note by Simpson (1758). Walmesley (1759) investigated the effects of the equatorial bulge of the Earth on the motion of the Moon, while (1762) was an attempt at a general theory of perturbations. Walmesley in his researches tried to vindicate Newton's results on perturbations against criticisms such as those of Clairaut.

By contrast, Thomas Simpson tried to produce new results. His (1757) is a remarkable text. Simpson correctly approximated the motion of the Moon's apogee by taking into consideration not only the component of the perturbative force of the Sun along the Earth–Moon radius vector, but also its tangential component. However, his claim of having attained this result independently from Clairaut (1752) hardly appears credible.

Instead of looking to the continent for inspiration, Matthew Stewart (1717–85) followed strictly geometrical methods. His preference for geometrical methods was certainly the result of the education he received in Glasgow, where in 1734 he became a student of Robert Simson. In 1741 he moved to Edinburgh where he attended Maclaurin's lectures. Stewart succeeded Maclaurin in the Chair of Mathematics in 1747 and retired from teaching in 1772. During this period he wrote (1746) and (1763a) on pure geometry, and (1761) and (1763b) on physical astronomy. His best achievements are definitely in the field of geometry, while his work on astronomy needs to be mentioned here because of the responses it evoked in Great Britain. Stewart's purpose was to tackle the study of planetary motions by using only geometrical methods. This project had also been contemplated by Simson.

Stewart's geometrical methods do not resemble those employed by Newton in the *Principia* or by Maclaurin in his (1742). His classicism led him to discard the geometrical approximations of the kinematical method of fluxions. He thought that it was possible to reinterpret the 'geometry of fluxions' within the framework of Archimedean geometry. The only mathematical tools permitted by him were well known from the times of the Greeks. His works are written without algebraical formulae; he prefers to write down an endless series of proportions.

Stewart's ambitious programme was to tackle the three-body problem with the aid of ancient geometry. Furthermore, he wanted to determine the distance of the Sun from the Earth by studying the effect of the disturbing force of the Sun on the motion of the Moon's apogee. His attempt would not have attracted much attention had he not been the Professor of Mathematics at the University of Edinburgh and one of the most reputed scientists in Scotland. Samuel Horsley, the editor of Newton's *Opera* (1779–85), praised Stewart's work in his papers in the *Philosophical Transactions*, (1768) and (1770). A first criticism appeared in Dawson's *Four Propositions* (1769). John Dawson (1734–1820) was merely a self-taught mathematician who worked as a surgeon and an itinerant schoolmaster, but he was able to indicate in more than 200 pages of analysis where the professor had gone wrong. Stewart's errors led him to give an estimation of the Sun's distance far superior to the one established by the observations of the transit of Venus in 1761. Another dart from a non-professional mathematician was fired by John Landen with his *Animadversions on Dr. Stewart's Computation of the Sun's Distance from the Earth* (1771b). Landen was able to dismiss Stewart's work in 18 pages of algebraical spot checks of the professor's 600 pages of geometrical proofs. However, it would be wrong to see Landen (1771b) as a successful criticism of the geometrical tradition of Newton's *Principia*, since Stewart's mathematics had little to do with the geometry of Newton. With Stewart a programme inaugurated by Simson in Glasgow and consisting in the development of the 'analysis of the Ancients' and its application to the study of natural philosophy came to an end. It had very little impact on the development of the calculus of fluxions and belongs more to the history of pure geometry.

The best studies on Newton's theory of the Moon are I. B. Cohen's introduction to Newton (1975), Waff (1976) and Whiteside (1976). In 1702 Newton published his revised theory (1702), which was published in Latin in D. Gregory (1702), pp. 332–6, and in English as a separate pamphlet. This small tract ran into four Latin printings and thirteen English printings in the eighteenth century. On Matthew Stewart see Playfair's biography, Playfair (1788a), reprinted in Playfair (1822a), IV, pp. 2–30.

2 On Clairaut see Greenberg (1979).
3 On Newton's, Bernoulli's and Euler's theories of tides see Aiton (1955).
4 See Todhunter (1873), I, pp. 133–75, for a modernized summary of Maclaurin's results.
5 Simpson reworked into a more orderly form these two essays in Simpson (1750c), II, pp. 445–79. I will follow this second version.
6 These theorems allow Simpson to study the form of equilibrium of a rotating fluid. He confines himself to studying Huygens's condition (only in the *Doctrine and Application of Fluxions* (1750c) is there a treatment of Newton's condition of balancing columns). His accurate study of the relationship between the eccentricity of a fluid oblatum and the angular velocity of rotation is a notable achievement. Simpson proves that 'if the angular velocity of rotation exceeds a certain limit, the oblatum is no longer a possible form of

equilibrium' (Todhunter (1873), I, p. 180), and he hints that below this value there are two oblata as forms of equilibrium. The latter result is suggested by a table in which the entries in the first column are proportional to the eccentricity, while those in the second are proportional to the corresponding angular velocity which meets the conditions of equilibrium. Since the entries in the first column are monotonically decreasing, while those in the second decrease to a minimum and then increase, it appears that for the same angular velocity there will be two forms of equilibrium. Simpson continues his essay by studying the case in which a spherical kernel is surrounded by a less dense fluid. In the second essay, which follows an analysis very similar to the first, Simpson treats the attraction at the surface of spheroids, which are not ellipsoids of revolution but are 'nearly spherical'. For further details see Todhunter (1873), I, pp. 176–88.

7 The original French reads as

j'ai jugé à propos de traiter en particulier de la figure des Spheroides homogenes, & d'abandonner ma Méthode, quant à ces Spheroides, pour suivre celle que M. Mac Laurin vient de donner dans son excellent Traité des Fluxions. Cette Méthode m'a paru si belle & si sçavante, que j'ai crû faire plaisir à mes Lecteurs de la mettre ici.

8 The original French reads as

Pour qu'un Spheroide fluide tournant autour de son Axe, & dans lequel la Loi de la gravité est donée, puisse conserver une forme constante ; il suffit qu'un Canal quelconque rentrant en lui-meme & placé dans le plan du Meridien de ce Spheroide, soit toujours en équilibre, en ne considérant que la seule force de la gravité sans la force centrifuge.

9 The original French reads as

Si on vouloit présentement faire usage de cette quantité, pour trouver en termes finis la valeur du poids du Canal ON, en supposant que la courbure de ce Canal fut donnée par une équation entre x & y, on commenceroit par faire évanouir y & dy de $Pdy + Qdx$; cette différentielle n'ayant plus que x & dx, on l'integreroit en observant de completter l'intégrale, c'est-à-dire d'ajouter la constante nécessaire, afin que le poids fut nul, lorsque x seroit égal à CG : On seroit ensuite $x = CI$, & l'on auroit le poids total de ON. Mais comme l'équilibre du Fluide demande que le poids de ON ne dépende pas de la courbure de OSN, c'est-à-dire de la valeur particulier de y en x, il faut donc que $Pdy + Qdx$ puisse s'intégrer sans connoitre la valeur de x, c'est-à-dire qu'*il faut que $Pdy + Qdx$ soit une différentielle complette, afin qu'il puisse y avoir équilibre dans le Fluide.*

10 The original French reads as

une quantité qui a pour intégrale une fonction de x & de y. $ydx + xdy$, $(ydx + xdy)/2\surd(aa + xy)$ sont des différentielles complettes, parce qu'elles ont pour intégrales xy, $\surd(a^2 + xy)$.

11 The original French reads as

la différentielle de la fonction P, prise en supposant x seulement variable.

Chapter 6

1 On the reluctance of the British to accept imaginary numbers, see Nagel (1935). Even an open minded mathematician like John Playfair could maintain in Playfair (1779) that the use of complex numbers was justified only in the case of the trigonometric functions because of the geometrical analogies between the circle and the hyperbola (see chapter 7, section 7.2).
2 From 1762 to 1788 Landen was land agent to William Wentworth, second Earl Fitzwilliam.
3 Some of Landen's letters to Simpson are reproduced in Clarke (1929), pp. 176–81.
4 I wish to acknowledge my indebtedness to Angelo Guerraggio for his comments on Landen's foundation of the calculus and Landen's work on elliptic integrals; see Guerraggio (1987).

5 An interesting anonymous article on Waring is in the additions to Hutton (1796, 1795), 2nd edn., pp. 717–26. I owe this reference to Ivor Grattan-Guinness.

6 Lagrange and Laplace were referred to in the second edition of Waring (1776).

7 The original Latin reads as

Data quantitate A, in qua contineatur duae variabiles quantitates x & y; sit ejus fluxio $\dot{A} = a\dot{x} + b\dot{y}$; inveniantur fluxiones quantitatum a & b, quae sint respective $\dot{a} = \alpha\dot{x} + \beta\dot{y}$, $\dot{b} = \pi\dot{x} + \rho\dot{y}$, ubi a, b, α, β, π, ρ, sunt functiones literarum x & y; tum erit $\pi = \beta$, i.e. π erit eadem quantitas ac β.

Cor. Hinc. data fluxione ($\dot{A} = a\dot{x} + b\dot{y}$) duas variabiles quantitates x & y involvente, inveniri potest; utrum ejus fluens exprimi potest, necne. Invenientur enim fluxiones quantitatum a & b, & si $\dot{a} = \alpha\dot{x} + \beta\dot{y}$, & $\dot{b} = \pi\dot{x} + \rho\dot{y}$, & $\pi = \beta$, tum exprimi potest fluens; sin aliter vero non.

8 The other equations are taken without recognition from Euler (1768–70). See Waring (1776), pp. 231–54, and 2nd edn., pp. 285–305.

Chapter 7

1 On the teaching of mathematics in the Scottish universities during the eighteenth century see Gibson (1927). On Glasgow University see Coutts (1909) and Mackie (1954). On Aberdeen University see Rait (1895) and Ponting (1979a) and (1979b); on King's College see P.J. Anderson (1893) and Innes (1854); on Marischal College see P.J. Anderson (1889–98). On St Andrews University see J.M. Anderson (1905), Dickinson (1952) and Cant (1970). On Edinburgh University see Grant (1884).

2 Trail studied at Marischal from 1759 to 1763. In 1766 he graduated B.A. at Glasgow. He wrote (1812) a biography of his friend Robert Simson.

3 At King's a professorship of mathematics was instituted in 1703. Thomas Bower (1660?–1724) was appointed in 1703, but his teaching was discontinued and he resigned in 1717. The second and last Professor of Mathematics at King's College in the eighteenth century was Alexander Rait. He was appointed in 1732.

4 On the Philosophical Society of Edinburgh see R. L. Emerson (1985).

5 On John Playfair see Jeffrey (1822).

6 However, a defence of Playfair (1779) appeared in a review of Woodhouse's paper (1801a). See anonymous (1803a).

7 Another important paper in which continental mathematics was compared with British mathematics is Toplis (1805).

8 On Ivory see the obituary in the *Proceedings of the Royal Society*, IV (1842), pp. 406–12.

9 On Wallace see Panteki (1987).

10 John Hellins (d.1827) published some papers in the *Philosophical Transactions* in the same years as Ivory and Wallace. He also concerned himself with the development of $(a^2 + b^2 - 2ab\cos(\theta))^n$. Hellins was a self-taught mathematician who worked for some time as an assistant of Maskelyne at the Greenwich Observatory. See Hellins (1781), (1788), (1794), (1796), (1798a), (1798b), (1800), (1802) and (1811).

11 The only biography of Spence is Galt (1820), prefaced to the posthumous Spence (1820).

12 Spence (1820) seems to be quite rare. Probably his friends supported the cost of the printing of a limited number of copies.

Chapter 8

1 Royal Warrant of 30 April 1741 quoted in W.D. Jones (1851), p. 1. W.D. Jones (1851) and Smyth (1961) are the best sources of information on Woolwich. See also Manners (1764), Townshend (1776), Dupin (1820) and Porter and Watson (1889–1915).

2 That is: David Gregory (1745) *A Treatise of Practical Geometry*; John Muller (1747) *The Attack and Defence of Fortify'd Places ...*; John Muller (1746) *A Treatise Containing the Elementary Part of Fortification ....* 'Vauban' is probably Cambray (1691) *The New Method of Fortification, as Practised by Monsieur de Vauban ...* (translated by A. Swall), a work

180 NOTES: PP. 110–20

published in French in 1689 at Amsterdam. Another possible candidate is Sébastien le Prestre de Vauban (1737, 1742) *De l'attaque et de la Defense des Places* (but there are no English translations).

3 That is: Nicholas Saunderson (1756b) *Select Parts of Professor Saunderson's Elements of Algebra* ...; Thomas Simpson (1760) *Elements of Geometry* ..., 2nd edn.; William Hawney (1717) *The Compleat Measurer* ...; John Joshua Kirby (1754) *Dr Brook Taylor's Method of Perspective made easy* ...; John Lodge Cowley (1765) *The Theory of Perspective Demonstrated* ...; Thomas Salmon (1749) *A New Geographical and Historical Grammar* ...; Joseph Harris (1731) *The Description and Use of the Globes and the Orrery* ....

4 On Charles Hutton see Bruce (1823) and O. Gregory (1823). A recent and very useful biography of Hutton is in Howson (1982), chapter 4, pp. 59–74. See also anonymous (1805).

5 Hutton (1821) appeared in the *Philosophical Transactions* just one year after Banks's death. David P. Miller sees behind the Hutton–Banks controversy an opposition between mathematicians and astronomers 'drawn from the middle and lower ranks of British provincial society' and the 'strong aristocratic flavour and natural historical bias of metropolitan scientific institutions' (Miller (1983), p. 10 *passim*). For a different opinion see Boas Hall (1984).

6 On the Royal Military College see Sandhurst (1802) and (1809), Smyth (1961) and Thomas (1961). Sandhurst (1849) and Mockler-Ferryman (1900) are less useful. See also the article 'College, Royal Military' in Hutton (1796, 1795), 2nd edn., the article 'Academy' in *Encyclopaedia Britannica*, supplement to the fourth, fifth and sixth editions and Dupin (1820).

7 Isaac Dalby (1744–1824) was Professor of Mathematics in the Senior Department of the Royal Military College from 1799 to 1820. Before being appointed, he had worked from 1787 to 1790 as an assistant to William Roy (1726–90), Major-General of the Royal Engineers, on the Trigonometrical Survey of England which began in 1784. The survey continued after 1791 under the direction of Colonel Williams and Captain Mudge. Williams (1762–1820) was a Woolwich man who later became Lieutenant-Governor of the Academy. Dalby published papers concerned with the survey in the *Philosophical Transactions*. Other details on Dalby's early career as a philomath can be gathered from his posthumous autobiography, Dalby (1830).

8 The *Mathematical Repository* is a quite rare publication. The best guide for the English mathematical serials is still Archibald (1929).

9 See Hutton (1775b), I, pp. 166–7, 258, 315, for the three pre-1730 fluxional answers by Anna Philomathes in 1719, Tapper in 1725 and Tho. Grant in 1729, respectively.

10 Wallace's translation of Legendre (1794) is in Leybourn (1806–35), II part 3 (1809), pp. 1–34, and III part 3 (1814), pp. 1–45. Wallace translated also in the first volume of the *Repository* (new series) two papers by Lagrange on spherical triangles and on numerical analysis (see Panteki (1987)).

11 Legendre (1794) was inserted 'at the request of several eminent mathematicians' (Leybourn (1806–35), II part 3 (1809), p. 1n).

12 The translation (Carnot (1800–01)) into English of Carnot's well-known work on the foundation of the calculus was by William Dickson and is in *Philosophical Magazine*, VIII (1800) and IX (1801).

13 The editions of the *Encyclopaedia Britannica* are as follows:

First edition (3 vols) Edinburgh, 1771 (published in parts from 1768). Another issue, London: *for* Dilly, 1773: another issue, London: *for* Donaldson, 1775.
Second edition (10 vols), Edinburgh, 1778–83; another issue, Dublin, 1791–7, plus Supplement, 1801.
Third edition (18+2 vols), Edinburgh 1797–1801.
Fourth edition (20 vols), Edinburgh, 1810.
Fifth edition (20 vols), Edinburgh, 1817.
Sixth edition (20 vols), Edinburgh, 1817.
Supplements to the fourth, fifth and sixth editions (6 vols), 1824.

Other encyclopaedias published before 1820 are: *Encyclopaedia Perthensis, Encyclopaedia*

Londiniensis, Encyclopaedia Mancuniensis and the *British Cyclopaedia*. For further information on this topic see Grattan-Guinness (1981) and (1985).

14 For further information on Wallace's articles see Panteki (1987).

15 Note that these encyclopaedias were often published in parts. The year on the title pages indicate when the work was finished. Each article was published some time before that date. For instance, the *Edinburgh Encyclopaedia* has 1830 on the title page, but the article 'Fluxions', which is in volume 9, was published in 1815. I owe this information to Ivor Grattan-Guinness and Maria Panteki.

16 Wallace, after reading Peacock (1834), wrote a letter to Peacock claiming priority in introducing the differential notation into Great Britain. This letter has been reproduced in Panteki (1987).

17 The attribution to Bonnycastle seems safe enough. This article is therefore quoted as Bonnycastle (1810). In the *Philosophical Magazine*, LVI (1820), pp. 218–24, there is a useful table of the authors who contributed to *The Cyclopaedia* (ed. A. Rees) (the papers appeared unsigned), as well as another chronological table of the time of publication of the articles. It appears that the article 'Functions' was published between 27 June and 8 October 1810. Possible authors include Peter Barlow, John Bonnycastle and John Pond, who are reported as contributors of articles on analysis and algebra. Ivory contributed articles on conic sections, curves and geometry.

18 For further information on the Royal Naval Academy see appendix D.3 and Lewis (1939), (1961) and (1965). I would like also to acknowledge my debt to Roger Bray (University of Essex) for his detailed letter on the Royal Naval Academy which I used in this section.

Chapter 9

1 The Senate House problems for the years 1801–10 were published in 1810 by a 'graduate of the University' in order to challenge the criticisms in Playfair (1808) against the study of mathematics in Cambridge. The anonymous editor asserted in the preface that Playfair's students would not have been able to pass the Cambridge exams. This is probably true. But, while Playfair had aimed to stimulate an interest in new methods and results in his lectures, the Cambridge students were required to attain a level of understanding in mathematics which did not go beyond the Cotesian forms. See anonymous (1810). George Atwood's researches are typical of a Cambridge educated mathematician. His rather elaborate works are notable for the history of engineering, but his mathematics is trivial: see Atwood (1784), (1794), (1796), (1798) and (1801).

2 On Vince see Grattan-Guinness (1986).

3 Nicholas Saunderson's lectures circulated widely in the eighteenth century. The manuscript copies I have seen are Cambridge University Library (Cambridge), MSS Add 6312, 2977, 589; University College (London), MSS Add 243; British Library (London) MS Add Eg. 834; Bodleian Library (Oxford) MS Rigaud 3–4.

4 The second edition of Dealtry (1810) was given a bad review, possibly by Playfair, in the *Edinburgh Review*: see ([Playfair] (1816)).

5 On Woodhouse see Becher (1980).

6 This appreciative review appeared anonymously in *The Monthly Review*. I quote it as [Woodhouse] (1799). Another review, probably written by Woodhouse, concerned Lacroix (1797–1800) large treatise on the calculus: see [Woodhouse] (1800).

7 Readers acquainted with Peacock's principle of permanence of equivalent forms will find many anticipations in Woodhouse's methodology. For Woodhouse's influence on Peacock and Babbage see Becher (1980).

8 We read in Wallis and Wallis (1986), p. 439, that Adam Walker (1731–1821), author of works on natural philosophy, book-keeping, the use of globes, and a well known scientific lecturer, taught in York in the 1760s and gave lectures for the Dublin Society.

9 On Dublin University see W.B.S. Taylor (1845).

10 On the reform of mathematics in Ireland see MacMillan (1984) and Grattan-Guinness (1988).

11 See Grattan-Guinness (1988) for a complete list of Irish works in the period from 1782 to 1840.

12 We can mention here L'Huilier (1796), a paper published in the *Philosophical Transactions* which might have inspired Woodhouse's formal approach to trigonometry. Thomas Knight was another British Lagrangian who published in the *Philosophical Transactions*. Knight (1811) and (1816) were concerned with Arbogast (1800); Knight (1817a), (1817b) and (1817c) were written in differential notation and referred to Spence (1809). See also Knight (1812a) and (1812b).

13 This famous pun is reported in Babbage (1864).

14 On the Analytical Society see Enros (1981) and (1983).

15 Peacock wrote:

M. Cauchy, in his *Leçons sur le Calcul Infinitésimal*, [...], has attempted to conciliate the direct consideration of *infinitesimals* with the purely algebraical view of the principles of this calculus, which Lagrange first securely established, [...]. He considers *all infinite series as fallacious which are not convergent*, [...], it must be an erroneous view of the principles of algebra which makes the result of any general operation dependent upon the fundamental laws of algebra to be fallacious. (Peacock (1834), pp. 247-8n)

16 John Herschel worked mainly on the calculus operators: see Herschel (1814), (1816), (1818) and (1822). Babbage devoted himself to the calculus of functions: see Babbage (1815), (1816), (1817a), (1817b) and (1822).

BIBLIOGRAPHY

Abbreviations used in the Bibliography are as follows

Phil. Trans.: Philosophical Transactions of the Royal Society;
Edin. Trans.: Transactions of the Royal Society of Edinburgh;
Irish Trans.: Transactions of the Royal Irish Academy.

Agnesi, Maria G. (1801) *Analytical Institutions, in Four Books: Originally Written in Italian, by Donna Maria Gaetana Agnesi, Professor of the Mathematics and Philosophy in the University of Bologna...*, (trans. John Colson; rev. John Hellins), London: *by* Taylor & Wilks

Aiton, Eric (1955) 'The contributions of Newton, Bernoulli, and Euler to the theory of tides', *Annals of Science*, XI, pp. 206–23

Anderson, James M. (ed.) (1905) *The Matriculation Roll of the University of St. Andrews 1747–1897*, Edinburgh and London: W. Blackwood

Anderson, Peter J. (ed.) (1889–98) *Fasti Academiae Mariscallanae Aberdonensis Selection from the Records of the Marischal College and University MDXCIII–MDCCCLX*, 3 vols, Aberdeen: *for the* New Spalding Club

Anderson, Peter J. (ed.) (1893) *Officers and Graduates of University & King's College Aberdeen MVD–MDCCCLX*, Aberdeen: *for the* New Spalding Club

anonymous (1724) 'An account of a book, intituled, Harmonia Mensurarum, sive Analysis & Synthesis per Rationum & Angulorum Mensuras Promotae: Accedunt Alia Opuscola Mathematica: per Rogerum Cotesium...', *Phil. Trans.*, XXXII (1722), pp. 139–50 (by Robert Smith?)

anonymous (1803a) (review of Woodhouse (1801a)), *The Edinburgh Review* (Jan. 1803), pp. 407–12

anonymous (1803b) 'A new method by W. Wallace', *The Edinburgh Review* (Jan. 1803), pp. 506–10

anonymous (1805) 'Biographical anecdotes of Charles Hutton, LLD FRS', *The Philosophical Magazine* (ed. Tilloch), XXI, pp. 62–7

anonymous (1810) *Cambridge Problems; Being a Collection of the Printed Questions Proposed to the Candidates for the Degrees of Bachelor of Arts, at the General Examination, From the Year 1801 to the Year 1810 Inclusive. With a Preface by a Graduate of the University*, Cambridge: *by* F. Hodson, *for* J. Deighton, *sold by* Longman *et al.*

anonymous (1849) *Complete Guide to the Junior and Senior Departments of the Royal Military College, Sandhurst...*, London: by C. Law

Arbogast, Louis F.A. (1800) *Du Calcul des Dérivations*, Strasbourg: by de Levrault

Archibald, Raymond C. (1929) 'Notes on some minor English mathematical serials', *Mathematical Gazette*, XIV, pp. 379–400

Atwood, George (1776) *A Description of the Experiments. Intended to Illustrate a Course of Lectures, in the Principles of Natural Philosophy. Read in the Observatory at Trinity College Cambridge*, London

Atwood, George (1784) *A Treatise on the Rectilinear Motion and Rotation of Bodies; with a Description of Original Experiments Relative to the Subject*, Cambridge: by J. Archdeacon, for J. & J. Merrill

Atwood, George (1794) 'Investigations, founded on the theory of motion, for determining the times of vibration of watch balances', *Phil. Trans.*, LXXXIV part 1 (1794), pp. 119–68

Atwood, George (1796) 'The construction and analysis of geometrical propositions, determining the positions assumed by homogeneal bodies which float freely, and at rest, on a fluid's surface; also determining the stability of ships, and of other floating bodies', *Phil. Trans.*, LXXXVI part 1 (1796), pp. 46–130

Atwood, George (1798) 'A disquisition on the stability of ships', *Phil. Trans.*, LXXXVIII part 2 (1798), pp. 201–308

Atwood, George (1801) *A Dissertation on the Construction and Properties of Arches*, London: by W. Bulmer, for Lunn and Egerton

Babbage, Charles (1815) 'An essay towards the calculus of functions', *Phil. Trans.*, CV part 2 (1815), pp. 389–423

Babbage, Charles (1816) 'An essay towards the calculus of functions', *Phil. Trans.*, CVI part 2 (1816), pp. 179–256

Babbage, Charles (1817a) 'Observations on the analogy which subsists between the calculus of functions and other branches of analysis', *Phil. Trans.*, CVII part 2 (1817), pp. 197–216

Babbage, Charles (1817b) 'Solutions of some problems by means of the calculus of functions', *The Journal of Sciences and the Arts, edited at the Royal Institution of Great Britain*, II, pp. 371–9

Babbage, Charles (1819) 'On some new methods of investigating the sums of several classes of infinite series', *Phil. Trans.*, CIX part 2 (1819), pp. 249–82

Babbage, Charles (1822) 'Observations on the notation employed in the calculus of functions, *Transactions of the Cambridge Philosophical Society*, I, pp. 63–76

Babbage, Charles (1823) 'On the application of analysis to the discovery of local theorems and porisms', *Edin. Trans.*, IX (1823), pp. 337–52

Babbage, Charles (1864) *Passages from the Life of a Philosopher...*, London: Longman et al.

[Babbage, Charles, and Herschel, John] (1813) *Memoirs of the Analytical Society*, Cambridge: by J. Smith

Babbage, Charles, Herschel, John, and Peacock, George (1820) *A Collection of Examples of the Applications of the Differential and Integral Calculus* [by Peacock], *A Collection of the Applications of the Calculus of Finite Differences* [by Herschel], *Examples of the Solutions of Functional Equations* [by Babbage], Cambridge: by J. Smith

Bailey, N., Gordon, G., Miller, P., and Lediard, T. (1736) *Dictionarium Britannicum. Or a more Compleat Universal Etymological English Dictionary than any Extant. Containing. [...] The second Edition with Numerous Additions and Improvements. By N. Bailey, Philologos. Assisted in the Mathematical Part by G. Gordon; in the Botanical by P. Miller; and in the Etymological, &c. by T. Lediard, Gent. Prof. of the Modern Languages in Lower Germany*, London: *for* T. Cox (1st edn. 1730)

Ball, William Rouse (1880) *The Origin and History of the Mathematical Tripos*, Cambridge University Press

Ball, William Rouse (1889) *A History of the Study of Mathematics at Cambridge*, Cambridge University Press

Barlow, Peter (1814a) *A New Mathematical and Philosophical Dictionary; comprizing an Explanation of the Terms and Principles of Pure and Mixed Mathematics, and such Branches of Natural Philosophy as are Susceptible of Mathematical Investigation. With Historical Sketches of the Rise, Progress, and Present State of the Several Departments of those Sciences. And an Account of the Discoveries and Writings of the Most Celebrated Authors, both Ancient and Modern*, London: *by* Whittingham and Rowland, *for* G. and S. Robinson et al.

Barlow, Peter (1814b) *New Mathematical Tables, Containing the Factors, Squares, Cubes, Square Roots, Cube Roots, Reciprocals, Hyperbolic Logarithms, of all Numbers from 1 to 10000; Tables of Powers and Prime Numbers; an Extensive Table of Formulae, or General Synopsis of the Most Important Particulars Relating to the Doctrine of Equations, Series, Fluxions, Fluents, &c. &c. &c.*, London: *for* G. and S. Robinson

Barlow, Peter (1817) *An essay on the Strength and Stress of Timber...*, London: *for* J. Taylor

Barlow, Peter (1820) *An Essay on Magnetic Attractions...*, London: for J. Taylor

Barrow, Isaac (1735) *Geometrical Lectures: Explaining the Generation, Nature and Properties of Curve Lines. Read in the University of Cambridge, by Isaac Barrow, D. D. Professor of Mathematicks, and Master of Trinity College, &c. Translated from the Latin Edition, Revised, Corrected and Amended by the Late Sir Isaac Newton by Edmund Stone FRS*, (trans. Edmund Stone), London: S. Austen

[Bayes, Thomas] (1736) *An Introduction to the Doctrine of Fluxions, and Defence of the Mathematicians Against the Objections of the Author of the Analyst, so far as they are Designed to Affect their General Methods of Reasoning*, London: *for* J. Noon

Bayes, Thomas (1764) 'A letter from the late Reverend Mr. Thomas Bayes, F.R.S. to John Canton, M.A. and F.R.S.', *Phil Trans.*, LIII (1763), pp. 269–71

Becher, Harvey (1980) 'Woodhouse, Babbage, Peacock, and Modern Algebra', *Historia Mathematica*, VII, pp. 389–400

[Berkeley, George] (1734) *The Analyst; or, a Discourse Addressed to an Infidel Mathematician. Wherein it is Examined Whether the Object, Principles, and Inferences of the Modern Analysts are more Distinctly Conceived, or more Evidently Deduced, than Religious Mysteries and Points of Faith, by the Author of the Minute Philosopher*, London: *for* J. Tonson (2nd edn. 1754: 3rd edn. in Berkeley (1784))

[Berkeley, George] (1735a) *A Defence of Free-Thinking in Mathematics. In Answer to a Pamphlet of Philalethes Cantabrigiensis, intituled, Geometry no Friend to Infidelity, or a Defence of Sir Isaac Newton, and the British Mathematicians. Also an Appendix Concerning Mr. Walton's Vindication of the Principles of Fluxions Against the*

Objections Contained in The Analyst. Wherein it is Attempted to put this Controversy in such a Light as that every Reader may be able to Judge thereof, Dublin: *by* M. Rhames, *for* R. Gunne (2nd edn, in Berkeley (1784))

[Berkeley, George] (1735b) *Reasons for not Replying to Mr. Walton's Full Answer in a Letter to P. T. P., By the Author of the Minute Philosopher*, Dublin: *by* M. Rhames, *for* R. Gunne (2nd edn. in Berkeley (1784))

Berkeley, George (1784) *The Works...*, Dublin: *by* J. Exshaw

[Blake, Francis] (1741) *An Explanation of Fluxions, in a Short Essay on the Theory*, London: *for* W. Innys (2nd edn. 1763; 3rd edn. 1809 in the 4th edn. of Rowe (1751))

Blake, Francis (1753) 'The best proportions for steam-engine cylinders, of a given content, consider'd', *Phil. Trans.*, XLVII (1751), pp. 197–201

Blake, Francis (1760) 'The greatest effect of engines with uniformly accelerated motion considered', *Phil. Trans.*, LI part I (1759), pp. 1–6

Boas Hall, Marie (1984) *All Scientists Now: The Royal Society in the Nineteenth Century*, Cambridge University Press

[Bonnycastle, John] (1810) 'Function, in analysis', in *The Cyclopaedia; or Universal Dictionary of Arts, Science, and Literature*, 39 + 6 vols, London: Longman, vol. XIV

Bonnycastle, John (1813) *A Treatise on Algebra, in Practice and Theory, in Two Volumes, with Notes and Illustrations; Containing a Variety of Particulars Relating to the Discoveries and Improvements that have been made in this Branch of Analysis*, 2 vols, London: J. Johnson & co. *et al.*

Bos, H. J. M. (1974) 'Differentials, higher-order differentials and the derivative in the Leibnizian calculus', *Archive for History of Exact Sciences*, XIV, pp. 1–90

Bowe, W. (1793) 'Some account of the life and writings of the author', in Emerson (1793), pp. i–xxii

Boyer, Carl B. (1959) *The History of the Calculus and its Conceptual Development*, New York: Dover

Brinkley, John (1800a) 'General demonstrations of the theorems for the sines and cosines of multiple circular arcs, and also of the theorems for expressing the powers of sines and cosines by the sines and cosines of multiple arcs; to which is added a theorem by help whereof the same method may be applied to demonstrate the properties of multiple hyperbolic areas', *Irish Trans.*, VII (1800), pp. 27–51

Brinkely, John (1800b) 'A general demonstration of the property of the circle discovered by Mr. Cotes, deduced from the circle only', *Irish Trans.*, VII (1800), pp. 151–9

Brinkley, John (1800c) 'A method of expressing, when possible, the value of one variable quantity in integral powers of another and constant quantities, having given equations expressing the relation of those variable quantities. In which is contained the general doctrine of reversion of series, of approximating to the roots of equations, and of the solution of fluxional equations by series', *Irish Trans.*, VII (1800), pp. 321–55

Brinkley, John (1802a) 'On the orbits in which bodies revolve, being acted upon by a centripetal force varying as any function of the distance, when those orbits have two apsides', *Irish. Trans.*, VIII (1802), pp. 215–31

Brinkley, John (1802b) 'On determining innumerable portions of a sphere, the solidities and spherical superficies of which portions are at the same time algebraically assignable', *Irish Trans.*, VIII (1802), pp. 513–25

Brinkley, John (1803a) 'An examination of various solutions of Kepler's problem, and a short practical solution of that problem, pointed out', *Irish Trans.*, IX (1803), pp. 83–131

Brinkley, John (1803b) 'A theorem for finding the surface of an oblique cylinder, with its geometrical demonstration, also, an appendix, containing some observations on the methods of finding the circumference of a very excentric ellipse; including a Geometrical demonstration of the remarkable property of elliptic areas discovered by Count Fagnani', *Irish Trans.*, IX (1803), pp. 145–58

Brinkley, John (1807) 'An investigation of a general term of an important series in the inverse method of finite differences', *Phil. Trans.*, XCVII part I (1807), pp. 114–32

Brinkley, John (1813) *Elements of Astronomy*, Dublin: *by* Graisberry and Campbell, printers to the university (chapters 1–16 in print 1808)

Brinkley, John (1820a) 'Investigations in physical astronomy, principally relative to the mean motion of the lunar perigree', *Irish Trans.*, XIII, pp. 25–51 (volume XIII has wrong title page 1818)

Brinkley, John (1820b) 'Observations relative to the form of the arbitrary constant quantities that occur in the integration of certain differential equations; and also in the integration of a certain equation of finite differences', *Irish Trans.*, XIII, pp. 53–61 (volume XIII has wrong title page 1818)

Brinkley, John (1820c) 'A method of correcting the approximate elements of the orbit of a comet, and the application of the same to observations made at the Observatory of Trinity College Dublin, on the comet of July 1819', *Irish Trans.*, XIII, pp. 189–98 (volume XIII has wrong title page 1818)

Brinkley, John (1825) 'The quantity of solar nutation as affecting the north polar distances of the fixed stars deduced from observation, and the application of this determination to confirm the conclusions relative to the parallaxes of certain fixed stars', *Irish Trans.*, XIV part 1 (1825), pp. 3–37

Bromhead, Edward F. (1816) 'On the fluents of irrational functions', *Phil. Trans.*, CVI part 2 (1816), pp. 335–54

Bruce, J. A. (1823) *A Memoir of Charles Hutton, LLD, FRS*, Newcastle: Hodgson

Cajori, Florian (1919) *A History of the Conceptions of Limits and Fluxions in Great Britain from Newton to Woodhouse*, Chicago and London: The Open Court

Cambray, Chevalier de (1691) *The New Method of Fortification, as Practised by Monsieur de Vauban*...(trans. A. Swall), London (2nd edn. 1693; 3rd edn. 1702; 4th edn. 1722; 5th edn. 1748)

Cant, R.G. (1970) *The University of St. Andrews; a Short History*, Edinburgh and London: Scottish Academic Press

Carnot, Lazare (1800, 1801) 'Reflections on the theory of the infinitesimal calculus...', *Philosophical Magazine* (ed. A. Tilloch), VIII, pp. 222–3, 224–40, 335–52; IX, pp. 39–55 (trans. William Dickson)

Carnot, Lazare (1803) *Géométrie de Position*, Paris: Duprat

Carré, Louis (1700) *Méthode pour la Mesure des Surfaces, la Dimension des Solides, leur Centres de Pesanteur, de Percussion et d'Oscillation par l'Application du Calcul Intégral*, Paris: J. Boudet

Chapman, Frederik Henrik (1820) *Treatise on Ship-building...*, (trans. James Inman), Cambridge: *by* J. Smith, *sold by* Deighton and Sons

Chasles, Michel (1875) *Aperçu Historique sur l'Origine et le Développement des Méthodes en Géométrie, Particulièrement de celles qui se Rapportent à la Géométrie Moderne*, 2nd edn., Paris: Gauthier-Villars (1st edn. 1837)

Cheyne, George (1703) *Fluxionum Methodus Inversa: sive Quantitatum Fluentium Leges Generaliores*, London: *by* J. Matthew, *sold by* R. Smith

Cheyne, George (1705) *Rudimentorum Methodi Fluxionum Inversae Specimina: quae Responsionem Continent ad Animadversiones Ab. de Moivre in Librum G. Cheynaei, M.D. S.R.S.*, London: *by* B. Motte, *for* G. Strahan

Cheyne, George (1715) *Philosophical Principles of Religion: Natural and Reveal'd. [...] Containing the Nature and Kinds of Infinites, their Arithmetick and Uses: together with the Philosophick Principles of Reveal'd Religion*, London: *for* G. Strahan

Clairaut, Alexis-Claude (1741a) 'Investigationes aliquot, ex quibus probetur terrae figuram secundum leges attractionis in ratione inversa quadrati distantiarum maxime ad ellipsin accedere debere', *Phil. Trans.*, XL (1737), pp. 19–25

Clairaut, Alexis-Claude (1741b) 'An inquiry concerning the figure of such planets as revolve about an axis, supposing the density continually to vary, from the center towards the surface', *Phil. Trans.*, XL (1738), pp. 277–306

Clairaut, Alexis-Claude (1743) *Théorie de la Figure de la Terre, tirée des Principes de l'Hydrostatique*, Paris: Durand

Clairaut, Alexis-Claude (1752) *Théorie de la lune deduite du seul principe de l'attraction réciproquement proportionelle aux quarrés des distances*, St. Petersburgh: Académie Impériale des Sciences

Clairaut, Alexis-Claude (1754) 'A translation and explanation of some articles of the book intitled, Théorie de la Figure de la Terre, by Mons. Clairaut, of the Royal Academy of Sciences at Paris, and FRS', *Phil. Trans.*, XLVIII part 1 (1753), pp. 73–85

Clark, Samuel, Croker, Henry and Williams, Thomas (1764, 1765, 1766) *The Complete Dictionary of Arts and Sciences. In which the Whole Circle of Human Learning is Explained. And the Difficulties Attending the Acquisition of Every Art, Whether Liberal or Mechanical, are Removed, In the Most Easy and Familiar Manner...*, 3 vols, London: *for* the authors, *sold by* J. Wilson *et al.*

Clarke, Frances M. (1929) *Thomas Simpson and his Times*, New York

Colden, Cadwallader (1751) 'An introduction to the doctrine of fluxions, or the arithmetic of infinities: in order to assist the imagination in forming conceptions of the principles on which that doctrine is founded' in *The Principles of Action in Matter, the Gravitation of Bodies, and the Motion of the Planets, Explained from those Principles*, pp. 189–215, London: *for* R. Dodsley

Collins, John (1713) *Commercium Epistolicum D. Johannis Collins et Aliorum de Analysi Promota: jussu Societatis Regiae In Lucem Editum*, London: *by* Pearson (2nd edn. 1722: 3rd edn. 1725)

Cotes, Roger (1717) 'Logometria', *Phil. Trans.*, XXIX (1714), pp. 5–45

Cotes, Roger (1722) *Harmonia Mensurarum, sive Analysis & Synthesis per Rationum*

& *Angulorum Mensuras Promotae: Accedunt Alia Opuscola Mathematica: per Rogerum Cotesium (ed & auxit R. Smith)*, (ed. and large additions by Robert Smith), [Cambridge University Press]

Cotes, Roger (1738) *Hydrostatical and Pneumatical Lectures*, London: *for the* editor, *sold by* S. Austen

Cotter, Charles H. (1968) *A History of Nautical Astronomy*, London, Sydney, Toronto: Hollis & Carter

Coutts, James (1909) *A History of the University of Glasgow from its Foundation in 1451 to 1909*, Glasgow: J. Maclehose and Sons

Cowley, John L. (1765) *The Theory of Perspective Demonstrated; in a Method Entirely New. By which Several Planes, Lines, and Points, Used in this Art, are Shewn by Moveable Schemes, in the True Positions in which they are Considered. Invented, and now Published for the Use of the Royal Academy at Woolwich*, London: *sold* J. Bennett

Craig(e), John (1685) *Methodus Figurarum Lineis Rectis & Curvis Comprehensarum Quadraturas Determinandi*, London: *for* M. Pitt

[Craig(e), John] (1688a) 'An account of a book. Methodus Figurarum Lineis Rectis & Curvis Comprehensarum Quadraturas Determinandi Authore J. Craige', *Phil. Trans.*, XVI (1686), p. 185

[Craig(e), John] (1688b) 'Additio ad Methodum figurarum quadraturas determinandi', *Phil. Trans.*, XVI (1686), pp. 186–9

Craig(e), John (1693) *Tractatus Mathematicus de Figurarum Curvilinearum Quadraturis et Locis Geometricis*, London: *by* S. Smith and B. Walford

Craig(e), John (1695) 'Tractatus Mathematicus de Figurarum Curvilinearum Quadraturis & Locis Geometricis. Authore Johanne Craig. London apud Sam. Smith & Benj Walford, Soc. Regiae Typographos', *Phil. Trans.*, XVIII (1694), pp. 113–14

Craig(e), John (1698a) 'De figurarum geometrice irrationalium quadraturis', *Phil. Trans.*, XIX (1697), pp. 708–11

Craig(e), John (1698b) 'Additio ad schedulam de quadraturis', *Phil. Trans.*, XIX (1697), pp. 785–7

Craig(e), John (1699) 'Quadratura logarithmicae', *Phil. Trans.*, XX (1698), pp. 373–4

Craig(e), John (1702) 'Epistola ad editorem continens solutionem duorum problematum', *Phil. Trans.*, XXII (1701), pp. 746–51

Craig(e), John (1704) 'Specimen methodi generalis determinandi figurarum quadraturas', *Phil. Trans.*, XXIII (1703), pp. 1346–60

Craig(e) John (1706) 'Solutio problematis, A Clariss. viro D. Jo. Bernoulli in Diario Gallico Febr. 1703 Propositi. Quam D. G. Cheyneo communicavit Jo. Craige', *Phil. Trans.*, XXIV (1704), pp. 1527–9

Craig(e), John (1710) 'De linearum curvarum longitudine', *Phil. Trans.*, XXVI (1708), pp. 64–6

Craig(e), John (1712) 'Logarithmotechnia generalis', *Phil. Trans.*, XXVII (1710), pp. 191–5

Craig(e), John (1718) *De Calculo Fluentium Libri duo. Quibus Subjunguntur Libri duo de Optica Analytica*, London: *ex* Officina Pearsoniana

Dalby, Isaac (1806) *A Course of Mathematics, Designed for the Use of the Officers and Cadets, of the Royal Military College*, 2 vols, London: *by* Glendinning, *for* the Author

Dalby, Isaac (1830) 'Mr. Dalby Late Professor of Mathematics in the Senior Department of the Royal Military College', in Leybourn (1806–35), pp. 196–203 [Dalby's autobiography]

Davies, Richard (1740) 'Memoirs of the life and character of Dr. Nicholas Saunderson', in Saunderson (1740), pp. i–xix

Dawson, John (1769) *Four Propositions, &c. Shewing, not only, that the Distance of the Sun, as Attempted to be Determined from the Theory of Gravity, by a Late Author, is, Upon his Own Principles, Erroneous; but also, that is more than Probable this Capital Question can Never be Satisfactorily Answered by any Calculus of the Kind*, Newcastle: *by* J. White & T. Saint, *for* W. Charnley

Dealtry, William (1810) *The Principles of Fluxions; Designed for the Use of Students in the University*, Cambridge: *by* J. Smith (2nd edn. 1816)

De Morgan, Augustus (1836) 'Calculus of functions', in *Encyclopaedia Metropolitana*, II, pp. 305–92 (date of volume II 1843)

De Morgan, Augustus (1852) 'On the early history of infinitesimals in England', *The London, Edinburgh and Dublin Philosophical Magazine and Journal of Science*, ser. IV, pp. 321–30

de Prony, Gaspard C.F.M.R. (1790–6) *Nouvelle Architecture Hydraulique...*, 2 vols, Paris: Firmin Didot

Desaguliers, John T. (1726a) 'A dissertation concerning the figure of the Earth', *Phil. Trans.*, XXXIII (1725), pp. 201–22, 239–55

Desaguliers, John T. (1726b) 'A dissertation concerning the figure of the Earth, part the second', *Phil. Trans.*, XXXIII (1725), pp. 277–304

Dickinson, William C. (1952) *Two Students at St. Andrews 1711–1716*, St. Andrews University Publications, 50, Edinburgh and London: Oliver and Boyd

Di Sieno, S., and Galuzzi, M. (1987) 'Calculus and geometry in Newton's mathematical work, some remarks', in S. Rossi (ed.) *Scienza e Immaginazione Nella Cultura Inglese del Settecento*, pp. 177–89, Milan: Unicopli

Ditton, Humphry (1704) 'De curvarum tangentibus e maximorum ac minimorum theoria immediate deductis; una cum theorem: quibusdam ad sectiones conicas pertinentibus, ejusdem calculi auxilio investigatis', *Phil. Trans.*, XXIII (1703), pp. 1333–45

Ditton, Humphry (1705) *The General Laws of Nature and Motion; with their Application to Mechanicks. Also the Doctrine of Centripetal Forces, and Velocities of Bodies, Describing any of the Conick Sections. Being a Part of the Great Mr. Newton's Principles. The Whole Illustrated with Variety of Useful Theorems and Problems, and Accommodated to the Use of the Younger Mathematicians*, London: *by* T. Mead, *for* J. Seller, C. Price and J. Senex

Ditton, Humphry (1706) *An Institution of Fluxions: Containing the First Principles, the Operations, with some of the Uses and Applications of that Admirable Method; According to the Scheme Perfix'd to his of Quadratures, by (its First Inventor) the Incomparable Sir Isaac Newton*, London: *by* W. Botham, *for* J. Knapton (2nd edn. 1726)

Dodson, James (ed.) (1748, 1753, 1755) *The Mathematical Repository*, 3 vols, London: *for* J. Nourse

Dupin, Charles (1820) *Force Militaire de la Grande-Bretagne*, in *Etudes et Travaux*, II, Paris: Bachelier

Eagles, Christina M. (1977a) *The Mathematical Work of David Gregory, 1659–1708*, Ph.D. thesis, Edinburgh University

Eagles, Christina M. (1977b) 'David Gregory and Newtonian science', *The British Journal for the History of Science*, X, part 3, pp. 216–25

Eames, John (1738) 'A brief account by Mr. John Eames F.R.S. of a work entitled, The Method of Fluxions and Infinite Series, with its Application to the Geometry of Curve Lines, by the inventor Sir Isaac Newton, Kt. &c. translated from the author's Latin original not yet made publick. To which is subjoin'd a perpetual comment upon the whole, &c. by John Colson', *Phil. Trans.*, XXXIX (1736), pp. 320–8

Eames, John (1741) 'An account by Mr. John Eames, FRS. of a book entituled A Mathematical Treatise, Containing a System of Conic-Sections, with the Doctrine of Fluxions and Fluents, Applied to Various Subjects. By John Muller', *Phil. Trans.*, XL (1737), pp. 87–9

Emerson, Roger L. (1985) 'The Philosophical Society of Edinburgh', *British Journal for the History of Science*, XVIII, pp. 255–303

Emerson, William (1743) *The Doctrine of Fluxions: not only Explaining the Elements thereof, but also its Application and Use in the Several Parts of Mathematics and Natural Philosophy*, London: *by* J. Bettenham, *sold by* W. Innys (2nd edn. corrected and enlarged 1757; 3rd edn. 1768; 4th edn. 1773)

Emerson, William (1754) *The Principles of Mechanics; Explaining and Demonstrating the General Laws of Motion, the Laws of Gravity, Motion of Descending Bodies, Projectiles, Mechanics Powers, Pendulums, Center of Gravity &c. Strength and Stress of Timber, Hydrostatics, and Construction of Machines*, London: *for* W. Innys and J. Richardson (2nd edn. 1758; 3rd edn. 1773; 4th edn. 1794; 5th edn. 1800; 6th edn. 1811; 7th edn. 1825; 8th edn. 1827; 9th edn. 1836)

Emerson, William (1763) *The Method of Increments. Wherein the Principles are Demonstrated; and the Practice Thereof Shewn in the Solution of Problems*, London: *for* J. Nourse (2nd edn. 1780)

Emerson, William (1763–91) *Cyclomathesis: Or an Easy Introduction to the Several Branches of the Mathematics. Being Principally Designed for the Introduction of Young Students, Before they Enter Upon the more Abstruse and Difficult Points Thereof*, 13 vols, London: *for* J. Nourse

Emerson, William (1767) *The Arithmetic of Infinites, and the Differential Method; Illustrated by Examples*, London: *for* J. Nourse (vol. V of Emerson (1763–91))

Emerson, William (1769a) *Mechanics; or the Doctrine of Motion...*, London: *for* J. Nourse (vol. VII of Emerson (1763–91), reprinted in Emerson (1793))

Emerson, William (1769b) *A System of Astronomy. Containing the Investigation and Demonstrations of the Elements of that Science*, London: *for* J. Nourse (vol. VIII of Emerson (1763–91))

Emerson, William (1793) *Tracts: Containing I. Mechanics, or the Doctrine of Motion. II. The Projection of the Sphere. III. The Laws of Centripetal and Centrifugal Force* [...]

a New Edition to which is prefixed Some Account of the Life and Writings of the Author, by the Rev. W. Bowe, London: *for* F. Wingrave

Engelsman, Steven B. (1984) *Families of Curves and the Origins of Partial Differentiation*, Amsterdam and Oxford: North Holland

Enros, Philip (1981) 'Cambridge University and the adoption of analytics in early nineteenth-century England', in H. Bos, H. Mehrtens and I. Schneider (eds.) *Social History of Nineteenth Century Mathematics*, pp. 135–47, Boston, Basel, Stuttgart: Birkhauser

Enros, Philip (1983) 'The Analytical Society (1812–1813): precursor of the renewal of Cambridge mathematics', *Historia Mathematica*, X, pp. 24–47

Euclid (1756) *The Elements of Euclid, viz. the First Six Books, Together with the Eleventh and Twelfth. In this Edition, the Errors, by which Theon, or Others, have long ago Vitiated these Books, are Corrected, and some of Euclid's Demonstrations are Restored* (ed. Robert Simson), Glasgow: R. and A. Foulis

Euler, Leonhard (1744) *Methodus Inveniendi Lineas Curvas Maximi Minimive Proprietates Gaudentes, sive Solutio Problematis Isoperimetrici Latissimo Sensu Accepti*, Lausanne and Geneve: *apud* Marcum-Michaelem Bousquet *& socios*

Euler, Leonhard (1768–70) *Istitutiones Calculi Integralis...*, 3 vols, Petropoli: *impensis* Academiae Imperialis Scientiarum

Euler, Leonhard (1776) *A Compleat Theory of the Construction and Properties of Vessels...*, (trans. Henry Watson), London: P. Elmsley

Fatio de Duillier, Nicolas (1699) *Lineae Brevissimi Descensus Investigatio Geometrica Duplex. Cui addita est Investigatio Geometrica Solidi Rotundi, in quod Minima fiat Resistentia*, London: *by* R. Everingham, *sold by* J. Taylor

Fatio de Duillier, Nicolas (1714) 'Epistola Nicolai Facis, [...] qua vendicat solutionem suam problematis de inveniendo solido rotundo seu tereti in quod minima fiat resistentia', *Phil. Trans.*, XXVIII (1713), pp. 172–6

Feather, John (1981) 'John Nourse and his authors', *Studies in Bibliography*, XXXIV, pp. 205–26

Feigenbaum, L. (1985) 'Brook Taylor and the Method of Increments', *Archive for History of Exact Sciences*, XXXIV, pp. 2–140

Fenn, Joseph (1769, 1772) *First Volume of the Instructions Given in the Drawing School Established by the Dublin-Society, Pursuant to their Resolution of the Fourth of February, 1768; To Enable Youth to Become Proficients in the Different Branches of that Art, and to Pursue with Success, Geographical, Nautical, Mechanical, Commercial, and Military Studies. Under the Direction of Joseph Fenn, Heretofore Professor of Philosophy in the University of Nants*, 2 vols, Dublin: *by* A. M'Culloch

Fleckenstein, Joachim O. (1956) *Der Prioritätstreit zwischen Leibniz und Newton*, Basel, Stuttgart: Birkhauser

Force, James E. (1985) *William Whiston: Honest Newtonian*, Cambridge University Press

Galt, John (1820) 'Biographical sketch of William Spence', in Spence (1820), pp. xvii–xxv

Gascoigne, John (1984) 'Mathematics and meritocracy: the emergence of the Cambridge Mathematical Tripos', *Social Studies in Science*, XIV, pp. 547–84

Gerard, Alexander (1755) *Plan of Education in the Marischal College and University of Aberdeen*, Aberdeen: *by* J. Chalmers

Gibson, G. A. (1927) 'Sketch of the history of mathematics in Scotland to the end of the 18th century', *Proceedings of the Edinburgh Mathematical Society*, Ser. 2, I (1927–9), pp. 1–18, 71–93

Giorello, Giulio (1985) *Lo Spettro e il Libertino: teologia, matematica, libero pensiero*, Milano: Mondadori

Glenie, James (1776) *The History of Gunnery, with a New Method of Deriving the Theory of Projectiles in Vacuo, From the Properties of the Square and Rombus*, Edinburgh: *for* J. Balfour, T. Cadell, J. Nourse

Gelnie, James (1778) 'The general mathematical laws which regulate and extend proportion universally, or, a method of comparing magnitudes of any kind together, in all the possible degrees of increase and decrease', *Phil. Trans.*, LXVII part 2 (1777), pp. 450–7

Glenie, James (1789) *The Doctrine of Universal Comparison, or General Proportion*, London: *by* J. Davis *for* G.G.J. and J. Robinson

Glenie, James (1793) *The Antecedental Calculus, or a Geometrical Method of Reasoning, Without any Consideration of Motion or Velocity Applicable to Every Purpose, to which Fluxions have been or can be Applied; with the Geometrical Principles of Increments, &c. and the Construction of some Problems as a few Examples Selected from an Endless and Indefinite Variety of them Respecting Solid Geometry, which he has by him in Manuscript*, London: *for* G.G.J. and J. Robinson

Glenie, James (1798) 'A short paper on the principles of the antecedental calculus', *Edin. Trans.*, IV part 2 (1798), pp. 65–82

Gompertz, Benjamin (1806) 'The application of a method of differences to the species of series whose sums are obtained by Mr. Landen, by the help of impossible quantities', *Phil. Trans.*, XCVI part 1 (1806), pp. 147–94

Gowing, Ronald (1983) *Roger Cotes – natural philosopher*, Cambridge University Press

Grant, Alexander (1884) *The Story of the University of Edinburgh, During its First Three Hundred Years*, 2 vols, London: Longmans and Green

Grattan-Guinness, Ivor (1969) 'Berkeley's criticism of the calculus as a study in the theory of limits', *Janus*, LVI, pp. 213–27

Grattan-Guinness, Ivor (1981) 'Mathematical physics in France, 1800–1840: knowledge, activity, and historiography', in J.W. Dauben (ed.) *Mathematical Perspectives...*, New York: Academic Press

Grattan-Guinness, Ivor (1985) 'Mathematics and mathematical physics from Cambridge, 1815–40: a survey of the achievements and of the French influences', in P.M. Harman (ed.) *Wranglers and Physicists, Studies on Cambridge Physics in the Nineteenth Century*, Manchester University Press

Grattan-Guinness, Ivor (1986) 'French "calcul" and English fluxions around 1800: some comparisons and contrasts', in *Jahrbuch Überblicke Mathematik*, pp. 167–78, Bibliographisches Institut Marginalien

Grattan-Guinness, Ivor (1988) 'Mathematical research and instruction in Ireland, 1782–1840', in John R. Nudds, Norman D. McMillan, Denis L. Weaize and

Susan McKenna Lawlor (eds.) *Science in Ireland 1800–1930: Tradition and Reform*, Dublin: Trinity College

Greenberg, John L. (1979) *Alexis Fontaine des Bertins, Alexis-Claude Clairaut, Integral Calculus, and the Earth's Shape, 1730–1743: an Episode in the History of the Role of Mathematics in the Rise of Science*, Ph.D. Thesis, University of Wisconsin-Madison

Greene, Robert (1707) *ΕΓΚΥΚΛΟΠΑΙΔΕΙΑ: or a Method of Instructing Pupils*, Cambridge (republished in Wordsworth (1877))

Gregory, David (1698a) 'De ratione temporis quo grave labitur per rectam data duo puncta conjungentem, ad tempus brevissimum quo, vi gravitatis, transit ab horum uno ad alterum per arcum cycloidis', *Phil. Trans.*, XIX (1697), pp. 424–5

Gregory, David (1698b) 'Davidis Gregorii MD Astronomiae Professoris Saviliani, & SRS, CATENARIA, ad Reverendum Virum D. Henricum Aldrich STP decanum Aedis Christi Oxoniae', *Phil. Trans.*, XIX (1697), pp. 637–52

Gregory, David (1700) 'Responsio ad animadversionem ad Davidis Gregorii Catenariam, Act. Eruditorum Lipsiae. Mense Februarii An 1699', *Phil. Trans.*, XXI (1699), pp. 419–26

Gregory, David (1702) *Astronomiae Physicae & Geometricae Elementa*, Oxford: e Theatro Sheldoniano (2nd English edn. by Edmund Stone 1726)

Gregory, David (1745) *A Treatise of Practical Geometry, in Three Parts*, (trans. and ed. Colin Maclaurin), Edinburgh: W. and T. Ruddimans

Gregory, Olynthus (1802) *A Treatise on Astronomy, in which the Elements of the Science are Deduced in a Natural Order, From the Appearances of the Heavens to an Observer on the Earth; Demonstrated on Mathematical Principles; and Explained by an Application to the Various Phaenomena*, London: for G. Kearsey (reissued in 1803)

Gregory, Olynthus (1806) *A Treatise of Mechanics; Theoretical, Practical and Descriptive*, 2 vols, London: by G. Kearsey

Gregory, Olynthus (1823) 'Brief memoir of the life and writings of Charles Hutton LLD...', *Imperial Magazine*, V, pp. 202–27

Guerraggio, Angelo (1987) 'Una fondazione algebrica del calcolo nell'Inghilterra del settecento', in S. Rossi (ed.) *Scienza e Immaginazione Nella Cultura Inglese del Settecento*, pp. 213–30, Milan: Unicopli

Hales, William (1778) *Sonorum Doctrina Rationalis, et Experimentalis, ex Newtoni Optimorumque Physicorum Scriptis, Methodo Elementaria Congesta. Cui praemittitur Disquisitio de Aere et Modificationibus Atmospherae*, Dublin: for G. Hallhead

Hales, William (1782) *De Motibus Planetarum in Orbibus Excentricis Secundum Theoriam Newtonianam. Dissertatio*, Dublin: excudebat J. Hill

Hales, William (1784) *Analysis Aequationum*, Dublin: excudebat J. Hill

Hales, William (1800) *Analysis Fluxionum*, London: J. Davis (republished in Maseres (ed.) (1791–1807), V, pp. 87–156)

Hall, Rupert (1980) *Philosophers at War. The Quarrel between Newton and Leibniz*, Cambridge University Press

Hamilton, Hugh (1758) *De Sectionibus Conicis Tractatus Geometricus. In quo, ex Natura ipsius Coni, Sectionum Affectioens Facillime Deducuntur Methodo Nova*, London: impensis W. Johnston

Hamilton, Hugh (1774) *Four Introductory Lectures in Natural Philosophy*, Dublin University Press, *reprinted for* John Nourse (1st edn. 1767)

Hamilton, Robert (1800) *Heads of Part of a Course of Mathematics [...] as Taught at Marischal College, Aberdeen*, Aberdeen

Hanna, John (1736) *Some Remarks on Mr. Walton's Appendix, which he Wrote in Reply to the Author of the Minute Philosopher; concerning Motion and Velocity*, Dublin: *by* S. Fuller, *for* the author

Hans, Nicholas (1951) *New Trends in Education in the Eighteenth Century*, London: Routledge & Kegan Paul

Harris, John (1702) *A New Short Treatise of Algebra: with the Geometrical Construction of Equations as far as the Fourth Power or Dimension. Together with a Specimen of the Nature and Algorithm of Fluxions.* London: *by* J. M. *for* D. Midwinter and T. Leigh (2nd edn. 1705; 3rd edn. 1714)

Harris, John (1704, 1710) *Lexicon Technicum: or, an Universal English Dictionary of Arts and Sciences: Explaining not only the Terms of the Art, but the Arts themselves*, 2 vols, London: *for* D. Brown, *et al.* (vol. I: 2nd edn. 1708, 3rd edn. 1716, 4th edn. 1725; vol II: 2nd edn. 1723; vols I and II in alphabetical order 1736, supplement 1744)

Harris, Joseph (1731) *The Description and Use of the Globes and the Orrery. To which is Prefixed, by way of Introduction, a Brief Account of the Solar System*, London: *for* T. Wright

Haüy, René J. (1807) *An Elementary Treatise on Natural Philosophy* (trans. O. Gregory), 2 vols, London: *for* G. Kearsey

Hawney, William (1717) *The Compleat Measurer: or, Whole Art of Measuring, in Two Parts...*, London: *for* D. Browne, *et al.*

Hayes, Charles (1704) *A Treatise of Fluxions: or, an Introduction to Mathematical Philosophy; Containing a Full Explication of that Method by which the most celebrated Geometers of the Present Age have made such Vast Advances in Mechanical Philosophy. A Work Very Useful for Those that Would Know How to Apply Mathematicks to Nature*, London: *by* E. Midwinter, *for* D. Midwinter and T. Leigh

Heath, Robert (1752) *Truth Triumphant: or, Fluxions for the Ladies. Shewing the Cause to be Before the Effect, and Different from it; That Space is not Speed, nor Magnitude Motion. With a Philosophic Vision, Most Humbly Dedicated to his Illustrious High, and Serene Excellence, the Sun. For the Information of the Public, by $\dot{X}$, $\dot{Y}$, and $\dot{Z}$, who are not of the Family of $\dot{x}$, $\dot{y}$ and $\dot{z}$, but near Relations of x', y' and z'*, London: *for* W. Owen

Hellins, John (1781) 'Theorems for computing logarithms', *Phil. Trans.*, LXX part 2 (1780), pp. 307–17

Hellins, John (1788) *Mathematical Essays on Several Subjects: Containing New Improvements and Discoveries in the Mathematics*, London: *for* the author, *sold by* L. Davis, J. Deighton, *et al.*

Hellins, John (1794) 'Dr. Halley's quadrature of the circle improved: being a transformation of his series for that purpose to others which converge by the powers of 80', *Phil. Trans.*, LXXXIV part 2 (1794), pp. 217–22

Hellins, John (1796) 'Mr. Jones's computation of the hyperbolic logarithm of 10 improved: being a transformation of the series which he used in that computation to others which converge by the powers of 80. To which is added a postscript, containing an improvement of Mr. Emerson's computation of the same logarithm', *Phil. Trans.*, LXXXVI part 1 (1796), pp. 135–41

Hellins, John (1798a) 'A new method of computing the value of a slowly converging series, of which all the terms are affirmative', *Phil. Trans.*, LXXXVIII part 1 (1798), pp. 183–99

Hellins, John (1798b) 'An improved solution of a problem in physical astronomy; by which, swiftly converging series are obtained, which are useful in computing the perturbations of the motions of the Earth, Mars and Venus, by their mutual attraction. To which is added an appendix, containing an easy method of obtaining the sums of many slowly converging series which arise in taking the fluents of binomial surds, &c.', *Phil. Trans.*, LXXXVIII part 2 (1798), pp. 527–66

Hellins, John (1800) 'A second appendix to the improved solution of a problem in physical astronomy, inserted in the Philosophical Transactions for the year 1798, containing some further remarks, and improved formulae for computing the coefficients A and B; by which the arithmetical work is considerably shortened and facilitated', *Phil. Trans.*, XC part 1 (1800), pp. 86–97

Hellins, John (1802) 'Of the rectification of the conic sections', *Phil. Trans.*, XCII part 2 (1802), pp. 448–76

Hellins, John (1811) 'On the rectification of the hyperbola by means of two ellipses...', *Phil. Trans.*, CI part 1 (1811), pp. 110–54

Helsham, Richard (1739) *A Course of Lectures in Natural Philosophy*, London: *published by* B. Robinson *for* J. Nourse

Herschel, John F.W. (1813) 'On a remarkable application of Cotes's theorem', *Phil. Trans.*, CIII part 1 (1813), pp. 8–26

Herschel, John F.W. (1814) 'Considerations of various points of analysis', *Phil. Trans.*, CIV part 2 (1814), pp. 440–68

Herschel, John. F.W. (1816) 'On the development of exponential functions, together with several new theorems relating to finite differences', *Phil. Trans.*, CVI part 1 (1816), pp. 25–45

Herschel, John F.W. (1818) 'On circulating functions, and on the integration of a class of equations of finite differences into which they enter as coefficients', *Phil. Trans.*, CVIII part 1 (1818), pp. 144–68

Herschel, John F.W. (1822) 'On the reduction of certain classes of functional equations to equations of finite differences', *Transactions of the Cambridge Philosophical Society*, I, pp. 77–88

Hodgson, James (1723) *A System of the Mathematics...*, London: *for* T. Page, William and Fisher Mount

Hodgson, James (1736) *The Doctrine of Fluxions, Founded on Sir Isaac Newton's Method, Published by Himself in his Tract upon the Quadrature of Curves*, London: *by* T. Wood *for* the author (2nd edn. 1756; 3rd edn. 1758)

Hofmann, Joseph E. (1943) 'Studien zur Vorgeschichte des Prioritätstreites zwischen Leibniz und Newton um die Entdeckung der höheren Analysis', in *Abhandlungen der Preussischen Akademie der Wissenschaften Jahrgang 1943. Math. – naturw. Klasse. Nr. 2*, Berlin: Akademie der Wissenschaften

Holliday, Francis (1745) *Syntagma Mathesios: Containing the Resolution of Equations:*

with a New Way of Solving Cubic and Biquadratic Equations, Analytically and Geometrically ..., London: *for* J. Fuller

Holliday, Francis (ed.) (1745–53) *Miscellanea Curiosa Mathematica*, 2 vols, London: E. Cave, *sold by* J. Fuller

Holliday, Francis (1777) *An Introduction to Fluxions, Designed for the Use, and Adapted to the Capacities of Beginners*, London: *for* J. Nourse (another 1821?)

Horsley, Samuel (1768) 'A computation of the distance of the Sun from the Earth', *Phil. Trans.*, LVII part 1 (1767), pp. 179–85

Horsley, Samuel (1770) 'On the computation of the Sun's distance from the Earth, by the theory of gravity', *Phil. Trans.*, LIX (1769), pp. 153–4

Horsley, Samuel (1779a) 'Logistica infinitorm', in Newton (1779–85), I, pp. 565–9

Horsley, Samuel (1779b) 'De geometria fluxionum liber singularis. Sive additamentum tractatus Newtoniani de methodo rationum primarum et ultimarum', in Newton (1779–85), I, pp. 573–92

Howson, Albert G. (1982) *A History of Mathematics Education in England*, Cambridge University Press

Hutton, Charles (1772) *The Principles of Bridges: Containing the Mathematical Demonstrations of the Properties of the Arches, the Thickness of the Piers, the Force of the Water against them, &c. Together with Practical Observations and Directions Drawn from the Whole*, Newcastle upon Tyne: *by* T. Saint

Hutton, Charles (ed.) (1775a) *Miscellanea Mathematica: Consisting of a Large Collection of Curious Mathematical Problems, and their Solutions. Together with Many Other Important Disquisitions in Various Branches of the Mathematics. Being the Literary Correspondence of Several Eminent Mathematicians*, London: G. Robinson, R. Baldwin (13 numbers 1771–5)

Hutton, Charles (ed.) (1775b) *The Diarian Miscellany: Consisting of All the Useful and Entertaining Parts, Both Mathematical and Poetical, Extracted from the Ladies Diary, from the Beginning of that Work in the Year 1704, Down to the End of the Year 1773. With many Additional Solutions and Improvements*, 5 vols, London: G. Robinson, R. Baldwin (published 1771–)

Hutton, Charles (1777) 'A new and general method of finding simple and quickly-converging series; by which the proportion of the diameter of the circle to its circumference may easily be computed to a great number of places of figures', *Phil. Trans.*, LXVI part 2 (1776), pp. 476–92

Hutton, Charles (1779a) 'The force of fired gun-powder, and the initial velocities of cannon balls, determined by experiments; from which is also deduced the relation of the initial velocity to the weight of the shot and the quantity of powder', *Phil. Trans.*, LXVIII part 1 (1778), pp. 50–85

Hutton, Charles (1779b) 'An account of the calculations made from the survey and measures taken at Schehallien, in order to ascertain the mean density of the Earth', *Phil. Trans.*, LXVIII part 2 (1778), pp. 689–788

Hutton, Charles (1780) 'Calculations to determine at what point in the side of a hill its attraction will be the greatest, &c.', *Phil. Trans.*, LXX part 1 (1780), pp. 1–14

Hutton, Charles (1781) 'Of cubic equations and infinite series', *Phil. Trans.*, LXX part 2 (1780), pp. 387–450

Hutton, Charles (1786) *Tracts, Mathematical and Philosophical*, London: G.G.J. and I. Robinson

Hutton, Charles (1790) 'Abstract of experiments made to determine the true resistance of the air to the surface of bodies, of various figures, and moved through it with different degrees of velocity', *Edin. Trans.*, II part 2 (1790), pp. 29–36

Hutton, Charles (1792) 'Memoirs of the life and writings of the author', in Simpson (1752), 2nd edn.

Hutton, Charles (1796, 1795) *A Mathematical and Philosophical Dictionary: Containing an Explanation of the Terms, and an Account of the Several Subjects, Comprized under the Heads Mathematics, Astronomy, and Philosophy both Natural and Experimental: with an Historical Account of the Rise, Progress, and Present State of these Sciences: also Memoirs of the Lives and Writings of the Most Eminent Authors, both Ancient and Modern, who by these Discoveries or Improvements have Contributed to the Advancement of them*, 2 vols, London: *by* J. Davies, *for* J. Johnson, J. Robinson (2nd edn, with additions and improvements 1815)

Hutton, Charles (1798, 1801) *A Course of Mathematics, in Two Volumes: Composed, and more Especially Designed, for the use of the Gentlemen Cadets in the Royal Military Academy at Woolwich*, 2 vols, London: G.G. and J. Robinson (3rd edn. 1800, 1801: 4th edn. 1803, 1804; 5th edn. 1806, 1807; 6th edn. 1810, 1811; in three volumes, vol. 3 with the assistance of Olynthus Gregory, another edn. 1813; 7th edn. 1819, 1820; several other edns. up to 1860)

Hutton, Charles (1812) *Tracts on Mathematical and Philosophical Subjects; Comprizing, among Numerous Important Articles, the Theory of Bridges, with Several Plans of Recent Improvements. Also the Result of Numerous Experiments on the Force of Gunpowder, with Applications to the Modern Practice of Artillery*, 3 vols, London: F.C. and J. Rivington

Hutton, Charles (1821) 'On the mean density of the Earth', *Phil. Trans.*, CI part 2 (1821), pp. 276–92

Huygens, Christiaan (1690) *Traité de la Lumiere, où sont Expliquées les Causes de se qui luy Arrive dans la Reflection, & dans la Refraction [...] avec un Discours de la Cause de la Pesanteur*, Leiden: Ches P. Vander

Inman, James (1810, 1812) *The System of Mathematical Education, at the Royal Naval College, Portsmouth*, 2 vols, Portsea: *Printed and sold by* W. Woodward

Inman, James (1829) *Nautical Tables...*, London: *for* C. and J. Rivington

Innes, C. (ed.) (1854) *Fasti Aberdonenses, selections from the Records of the University and King's College of Aberdeen 1494–1854*, Aberdeen: *for* the Spalding Club

Ivory, James (1798) 'A new series for the rectification of the ellipsis; together with some observations on the evolution of the formula $(a^2 + b^2 - 2ab \cos \theta)^n$', *Edin. Trans.*, IV part 2 (1798), pp. 177–90

Ivory, James (1805) 'A new and universal solution of Kepler's problem', *Edin. Trans.*, V part 2 (1805), pp. 203–46

Ivory, James (1809) 'On the attractions of homogeneous ellipsoids', *Phil. Trans.*, XCIX part 2 (1809), pp. 345–72

Ivory, James (1812a) 'On the grounds of the method which Laplace has given in the second chapter of the third book of his Mécanique Céleste for computing the attractions of spheroids of every description', *Phil. Trans.*, CII part 1 (1812), pp. 1–45

Ivory, James (1812b) 'On the attraction of an extensive class of spheroids', *Phil. Trans.*, CII part 1 (1812), pp. 46–82

Ivory, James (1814) 'A new method of deducing a first approximation to the orbit of a comet from three geocentric observations', *Phil. Trans.*, CIV part 1 (1814), pp. 121–86

Ivory, James (1822) 'On the expansion in a series of the attraction of a spheroid', *Phil. Trans.*, CXII part 1 (1822), pp. 99–112

Ivory, James (1823) 'On the astronomical refractions', *Phil. Trans.*, CXIII part 2 (1823), pp. 409–25

Jeffrey, Francis (1822) 'Biographical account of the late Professor Playfair', in Playfair (1822a), I, pp. xi–lxxvi

Jones, William (1706) *Synopsis Palmariorum Matheseos; or, a New Introduction to the Mathematics: Containing the Principles of Arithmetic & Geometry Demonstrated, in a Short and Easie Method, with their Applications to the Most Useful Parts Thereof [...] Design'd for the Benefit, and Adapted to the Capacities of Beginners*, London: by J. Matthews, for J. Wale

Jones, William (1772) 'Of logarithms', *Phil. Trans.*, LXI part 2 (1771), pp. 455–61

Jones, William D. (ed.) (1851) *Records of the Royal Military Academy*, Woolwich: The Royal Artillery Institution

[Jurin, James] (1734) *Geometry no Friend to Infidelity: or, a Defence of Sir Isaac Newton and the British Mathematicians, in a Letter to the Author of the Analyst Wherein it is Examined how far the Conduct of such Divines as Intermix the Interest of Religion with their Private Disputes and Passions, and Allow Neither Learning nor Reason to those they differ from, is of Honour or Service to Christianity, or Agreeable to the Example of our Blessed Saviour and his Apostles*, London: for T. Cooper (pseudonym used by the author: Philalethes Cantabrigiensis)

[Jurin, James] (1735) *The Minute Mathematician; or, The Free-Thinker no Just-Thinker. Set forth in a Second Letter to the Author of the Analyst; Containing a Defence of Sir Isaac Newton and the British Mathematicians, against a late Pamphlet, entituled, A Defence of Free-Thinking in Mathematicks*, London: for T. Cooper (pseudonym used by the author: Philalethes Cantabrigiensis)

Jurin, James (1744a) 'De mensura & motu aquarum fluentium, tentamen primum', *Phil. Trans.*, XLI part 1 (1739), pp. 5–40

Jurin, James (1744b) 'Tentaminis de mensura et motu aquarum fluentium, praecedente transactionum numero communicati, pars reliqua', *Phil. Trans.*, XLI part 1 (1739), pp. 65–91

Keill, John (1702) *Introductio ad Veram Physicam, seu Lectiones Physicae, Habitae in Schola Naturalis Philosophiae Academiae Oxoniensis. Quibus accedunt C. Hugenii theoremata de vi centrifuga et Motu Circulari Demonstrata*, Oxford: e Theatro Sheldoniano

Keill, John (1710) 'Jo. Keill ex Aede Christi Oxoniensis, A.M. Epistola ad Clarissimum Virum Edmundum Halleium Geometriae Professorem Savilianum, de legibus virium centripetarum', *Phil. Trans.*, XXVI (1708), pp. 174–88

Keill, John (1714) 'Problematis Kepleriani, de inveniendo vero motu planetarum, areas tempori proportionales in orbibus ellipticis circa focorum alterum describentium, solutio Newtoniana; à D. J. Keill, Astr. Prof. Savil. Oxon. & R.S.S. demonstrata & exemplis illustrata', *Phil. Trans.*, XXVIII (1713), pp. 1–10

Keill, John (1717a) 'Theoremata quaedam infinitam materiae divisibilitatem spectantia, quae ejusdem raritatem & tenuem compositionem demonstrant, quorum ope plurimae in physica tolluntur difficultates', *Phil. Trans.*, XXIX (1714), pp. 82–6

Keill, John (1717b) 'Observationes in ea quae edidit celeberrimus geometra Iohannes Bernoulli in Commentariis Physico mathematicis Parisiensibus anno 1710. De inverso problemate virium centripetarum. Et ejusdem problematis solutio nova', *Phil. Trans.*, XXIX (1714), pp. 91–111

Keill, John (1718) *Introductio ad veram Astronomiam, seu Lectiones Astronomicae Habitae in Schola Astronomica Academiae Oxoniensis*, Oxford: e Theatro Sheldoniano, *impensis* H. Clements

Keill, John (1720) *An Introduction to Natural Philosophy: or, Philosophical Lectures Read in the University of Oxford, Anno Dom. 1700. To which are Added the Demonstrations of Monsieur Huygens's Theorems, Concerning the Centrifugal Force and Circular Motion*, London: by H.W., for H.W. & J. Innys and J. Osborn (1st English translation of the 3rd edn. of Keill (1702))

Kirby, John J. (1754) *Dr Brook Taylor's Method of Perspective Made Easy, Both in Theory and Practice. In Two Books...*, Ipswich: by W. Craighton, for the author

Kirkby, John (1748) *The Doctrine of Ultimators. Containing a New Acquisition to Mathematical Literature, Naturally Resulting from the Consideration of an Equation, as Reducible from its Variable to its Ultimate State: Or, a Discovery of the True and Genuine Foundation of what has Hitherto Mistakenly Prevailed under the Improper Names of Fluxions and the Differential Calculus. By means of which we now have that Area of all Mathematical Science Entirely Rescued from the Blind and Ingeometrical Method of Deduction, which it has Hitherto Laboured Under; and made to Depend upon Principles as Strictly Demonstrable, Most Self-evident Proposition in the First Elements of Geometry*, London: for J. Hodges

Kitcher, Peter (1973) 'Fluxions, limits, and infinite littlenesse: a Study of Newton's presentation of the calculus', *Isis*, LXIV, pp. 33–49

Klingenstierna, Samuel (1733) 'Curvarum hyperbolicarum, aequationibus trium nominum utcunque definitarum, quadratura generalis duplici theoremate exhibita', *Phil. Trans.*, XXXVII (1731), pp. 45–50

Knight, Thomas (1811) 'On the expansion of any functions of multinomials', *Phil. Trans.*, CI part 1 (1811), pp. 49–88

Knight, Thomas (1812a) 'On the attraction of such solids as are terminated by planes; and of solids of greatest attraction', *Phil. Trans.*, CII part 2 (1812), pp. 247–309

Knight, Thomas (1812b) 'Of the penetration of an hemisphere by an indefinite number of equal and similar cylinders', *Phil. Trans.*, CII part 2 (1812), pp. 310–13

Knight, Thomas (1816) 'A new demonstration of the binomial theorem', *Phil. Trans.*, CVI part 2 (1816), pp. 331–4

Knight, Thomas (1817a) 'Of the construction of logarithmic tables', *Phil. Trans.*, CVII part 2 (1817), pp. 217–33

Knight, Thomas (1817b) 'Two general propositions in the method of differences', *Phil. Trans.*, CVII part 2 (1817), pp. 234–44

Knight, Thomas (1817c) 'Note respecting the demonstration of the binomial theorem inserted in the last volume', *Phil. Trans.*, CVII part 2 (1817), pp. 245–51

Koppelman, Elaine (1971) 'The calculus of operations and the rise of abstract algebra', *Archive for History of Exact Sciences*, VIII, pp. 155–242

Krieger, H. (1968) 'Uber Stirlings Bestimmung von $\sum_{n=0}^{\infty} (-1)^n/2n+1$ und $\sum_{n=1}^{\infty} 1/n^2$'. *Archive for History of Exact Sciences*, V, pp. 37–46

Lacroix, Silvestre F. (1797–1800) *Traité du Calcul Différentiel et du Calcul Intégral*, 3 vols, Paris: *chez* J.B.M. Duprat

Lacroix, Silvestre F. (1802) *Traité Elémentaire de Calcul Différentiel et de Calcul Intégral...*, Paris: *chez* Duprat

Lacroix, Silvestre F. (1816) *An Elementary Treatise of the Differential and Integral Calculus*, Cambridge: *by* J. Smith *for* J. Deighton (trans. of 2nd edn. of Lacroix (1802) by C. Babbage, J. Herschel and G. Peacock)

Lagrange, Joseph-Louis (1788) *Méchanique Analitique...*, Paris: *chez la* Veuve Desaint

Lagrange, Joseph-Louis (1797) *Théorie des Fonctions Analytiques...*, Paris: Imprimerie de la République

Landen, John (1755a) 'An investigation of some theorems which suggest some remarkable properties of the circle, and are of use in resolving fractions, whose denominators are certain multinomials, into more simple ones', *Phil. Trans.*, XLVIII part 2 (1754), pp. 566–78

Landen, John (1755b) *Mathematical Lucubrations: Containing New Improvements in the Various Branches of the Mathematics*, London: *for* J. Nourse

Landen, John (1758) *A Discourse Concerning the Residual Analysis: A New Branch of the Algebraic Art, Of Very Extensive Use, Both in Pure Mathematics and Natural Philosophy*, London: *for* J. Nourse

Landen, John (1761) 'A new method of computing the sums of certain series', *Phil. Trans.*, LI part 2 (1760), pp. 553–65

Landen, John (1764) *The Residual Analysis; a New Branch of the Algebraic Art, Of Very Extensive Use, Both in Pure Mathematics, and Natural Philosophy. Book I*, London: *for* the author

Landen, John (1769) 'A specimen of a new method of comparing curvilinear areas; by which many such areas may be compared as have not yet appeared to be comparable by any other method', *Phil. Trans.*, LVIII (1768), pp. 174–80

Landen, John (1771a) 'Some new theorems for computing the areas of certain curve lines', *Phil. Trans.*, LX (1770), pp. 441–3

Landen, John (1771b) *Animadversions on Dr. Stewart's Computation of the Sun's Distance from the Earth*, London: *by* G. Bigg, *for* the Author, *sold by* J. Nourse

Landen, John (1772) 'A disquisition concerning certain fluents, which are assignable by the arcs of the conic sections; wherein are investigated some new and useful theorems for computing such fluents', *Phil. Trans.*, LXI part 1 (1771), pp. 298–309

Landen, John (1775) 'An investigation of a general theorem for finding the length of any arc of any conic hyperbola, by means of two elliptic arcs, with some other new and useful theorems deduced therefrom', *Phil. Trans.*, LXV part 2 (1775), pp. 283–9

Landen, John (1777) 'A new theory of the rotatory motion of bodies affected by forces disturbing such motion', *Phil. Trans.*, LXVII part 1 (1777), pp. 266–95

Landen, John (1780, 1789) *Mathematical Memoirs Respecting a Variety of Subjects...,* 2 vols, London: vol. 1 *for the Author, sold by* J. Nourse; vol. 2 *for the Author, sold by* F. Wingrave

Landen, John (1781) *Observations on Converging Series, Occasioned by Mr. Clarke's Translation of Mr. Lorgna's Treatise on the Same Subject,* London: *for the author, sold by* J. Nourse

Landen, John (1783) *An Appendix to Observations on Converging Series,* London: J. Nourse

Landen, John (1784) *A Supplement to Observations on Converging Series,* London: J. Nourse

Landen, John (1785) 'Of the rotatory motion of a body of any form whatever, revolving, without restraint, about any axis passing through its center of gravity', *Phil. Trans.*, LXXV part 2 (1785), pp. 311–32

Landerbeck, Nicolao (1784) 'Methodus inveniendi lineas curvas ex proprietatibus variationis curvaturae', *Phil. Trans.*, LXXIII part 2 (1783), pp. 456–73; LXXIV part 2 (1784), pp. 477–500

Laplace, Pierre-Simon Marquis de (1814) *Treatise upon Analytical Mechanics; being the First Book of the Mechanique Celeste of P. S. Laplace* (trans. and notes by John Toplis), Nottingham: H. Barnett

Lardner, Dionysius (1825) *An Elementary Treatise on the Differential and Integral Calculus,* London: *for* J. Taylor

Lawrence, Paul D. (1971) *The Gregory Family: A Biographical and Bibliographical Study. To which is Annexed a Bibliography of the Scientific and Medical Books in the Gregory Library, Aberdeen University Library,* Ph.D. Thesis, University of Aberdeen

Lax, William (1799) 'A method of finding the latitude of a place, by means of two altitudes of the Sun and the time elapsed betwixt the observations', *Phil. Trans.*, LXXXIX part 1 (1799), pp. 74–120

Legendre, Adrien-Marie (1794) 'Mémoire sur les transcendantes élliptiques', *Mémoires de l'Académie Royale des Sciences,* pp. 1–102

Leibniz, Gottfried Wilhelm von (1684) 'Nova methodus pro maximis et minimis itemque tangentibus, quae nec fractas nec irrationales quantitates moratur et singulare pro illis calculi genus', *Acta Eruditorum* (October 1684), pp. 467–73 (republished in *Leibnizens mathematische Schriften,* V, pp. 220–6)

Lewis, Michael (1939) *England's Sea-Officers: The Story of the Naval Profession,* London: George Allen & Unwin

Lewis, Michael (1961) *A Social History of the Navy, 1793–1815,* London: George Allen & Unwin

Lewis, Michael (1965) *The Navy in Transition, 1814–1864, A Social History,* London: Hodder and Stoughton

Leybourn, Thomas (ed.) (1795–1804) *The Mathematical Repository: Containing*

Many Ingenious and Useful Essays and Extracts, with a Collection of Problems and Solutions, Selected from the Correspondence of Several Able Mathematicians, and the Works of Those who are Eminent in Mathematics, 3 vols, London: *for* the editor, *sold by* Allen and West and Glendinning (published in numbers 1795– with the title *The Mathematical and Philosophical Repository*, nos 1–8; *The Mathematical and Philosophical Repository, and Review*, nos. 9–14, and then bound in three volumes)

Leybourn, Thomas (ed.) (1806–35) *New Series of the Mathematical Repository*, 6 vols, London: *Printed and sold by* W. Glendinning (published in numbers 1804–, bound in six volumes published in 1806, 1809, 1814, 1819, 1830, 1835)

Leybourn, Thomas (ed.) (1817) *The Mathematical Question, Proposed in the Ladies' Diary...*, 4 vols, London: *by* W. Glendinning, *et al.*

L'Hospital, Guillaume-François-Antoine de (1696) *Analyse des Infiniment Petits, pour l'Intelligence des Lignes Courbes*, Paris: Imprimerie Royale

L'Huilier, Simon (1796) 'Manière élémentaire d'obtenir les suites par lesquelles s'expriment les quantités exponentielles et les fonctions trigonométriques des arcs circulaires', *Phil. Trans.*, LXXXVI part 1 (1796), pp. 142–65

Ludlam, William (1770) *Two Mathematical Essays. The First on Ultimate Ratios, the Second on the Power of the Wedge*, Cambridge: J. Archdeacon

Lyons, Israel (1758) *A Treatise of Fluxions*, London: W. Bowyer, *sold by* A. Millar, B. Dodd, J. Fletcher

Lyons, Israel (1775) 'Calculations in spherical trigonometry abridged', *Phil. Trans.*, LXV part 2 (1775), pp. 470–83

Machin, John (1720) 'Inventio curvae quam corpus descendens brevissimo tempore describeret; urgente vi centripeta ad datum punctum tendente, quae crescat vel decrescat juxta quamvis potentiam distantiae a centro; dato nempe imo curvae puncto & altitudine in principio casus', *Phil. Trans.*, XXX (1718), pp. 860–2

Machin, John (1726) 'De motu nodorum Lunae', in Newton (1687), 3rd edn., book III, pp. 451–4

Machin, John (1729) *The Laws of the Moon's Motion According to Gravity*, in Newton (1687), 1st English 'Motte' edn. 1729

Machin, John (1741) 'The solution of Kepler's problem', *Phil. Trans.*, XL (1738), pp. 205–30

Mackie, J.D. (1954) *The University of Glasgow 1451–1951, A Short History*, Glasgow: Jackson, sons and co.

Maclaurin, Colin (1720a) 'Tractatus de curvarum constructione & mensura; ubi plurimae series curvarum infinitae vel rectis mensurantur vel ad simpliciores curvas reducuntur', *Phil. Trans.*, XXX (1718), pp. 803–12

Maclaurin, Colin (1720b) 'Nova methodus universalis curvas omnes cujuscunque ordinis mechanicae describendi sola datorum angulorum & rectarum ope', *Phil. Trans.*, XXX (1719), pp. 939–45

Maclaurin, Colin (1720c) *Geometria Organica: sive Descriptio Linearum Curvarum Universalis*, London: *by* W. and J. Innys

Maclaurin, Colin (1724) 'Demonstration des Loix du choc des corps', *Piece qui a remporté le Prix de l'Academie Royale des Sciences, Proposé pour l'année mil sept cens vingt-quatre...*, Paris: C. Jombert

Maclaurin, Colin (1741) 'De causa physica fluxus et refluxus maris', in *Pieces qui ont remporté le Prix de l'Académie Royale des Sciences, en M.DDC.XL. Sur le Flux & Reflux de la Mer...*, Paris: G. Martin, J. B. Coignard, & les Freres Guerin

Maclaurin, Colin (1742) *A Treatise of Fluxions, in Two Books*, Edinburgh: T. W. and T. Ruddimans (2nd edn. 1801; French edn. 1749; *Abregé du Calcul Integral*, French abridged edn., 1765)

Maclaurin, Colin (1744a) 'A rule for finding the meridional parts to any spheroid, with the same exactness as in a sphere', *Phil. Trans.*, XLI part 2 (1741), pp. 808–9

[Maclaurin, Colin] (1744b) 'An account of a book intituled, A Treatise of Fluxions, in two books...', *Phil. Trans.*, XLII (1743), pp. 325–63

[Maclaurin, Colin] (1744c) 'The continuation of an account of a Treatise of Fluxions, &c. Book II', *Phil. Trans.*, XLII (1743), pp. 403–15

Maclaurin, Colin (1748a) *An Account of Sir Isaac Newton's Philosophical Discoveries, in four Books*, (ed. Patrick Murdoch), London: *for the author's children, sold by* Millar, *et al.*

Maclaurin, Colin (1748b) *A Treatise of Algebra, in three Parts. Containing I. The Fundamental Rules and Operations II. The Composition and Resolution of Equations of all Degrees; and the different Affections of their Roots III. The Application of Algebra and Geometry to each other. To which is added an Appendix, Concerning the General Properties of Geometrical Lines*, London: *for* A. Millar, J. Nourse

Maclaurin, Colin (1754a) 'Of the cause of the variation of the obliquity of the ecliptic', *Essays and Observations Physical and Literary, Read before a Society in Edinburgh, and published by them*, I, pp. 173–83

Maclaurin, Colin (1754b) 'Concerning the sudden and surprizing changes observed in the surface of Jupiter's body', *Essays and Observations Physical and Literary, Read before a Society in Edinburgh, and published by them*, I, pp. 184–8

Maclaurin, Colin (1982) *The Collected Letters of Colin Mac Laurin* (ed. S. Mills), Nantwich: Shiva Publishing

MacMillan, N.D. (1984) 'The analytical reform of Irish mathematics 1800–1831; *Newsletter of the Irish Mathematical Society*, no. 10, pp. 61–75

Manners, John (1764) *Rules and Orders for the Royal Military Academy at Woolwich*, London: *by* J. Bullock, J. Spencer and J. Bullock Junior

Martin, Benjamin (1736) *The Young Student's Memorial Book, or Pocket Library: Containing [...] A Very Large Collection of Theorems and Canons for Solving Questions and Problems in the Various Parts of Arithmetic, Algebra and Fluxions...*, London: *for* J. Noon

Martin, Benjamin (1739) Παυγεωμετρια; *or the Elements of all Geometry. containing [...] An Appendix, containing an Epitome of the Doctrine of Fluxions; and a Specimen of the Method de Maximis & Minimis...*, London: *for* J. Noon

Martin, Benjamin (1759) 'The doctrine of fluxions or elements of the new geometry', in *A New and Comprehensive System of Mathematical Institutions, Agreeable to the Present State of the Newtonian Mathesis...*, I, pp. 361–410, London: *printed and sold by* W. Owen

Martin, Benjamin (1773) *The Young Trigonometer's New Guide; Containing [...] Fluxionary Trigonometry...*, London: *for the author*

Maseres, Francis (1777) 'A method of finding the value of an infinite series of decreasing quantities of a certain form, when it converges too slowly to be summed in the common way by the mere computation and addition or subtraction of some of its initial terms', *Phil. Trans.*, LXVII part 1 (1777), pp. 187–230

Maseres, Francis (1779) 'A method of finding, by the help of Sir Isaac Newton's binomial theorem, a near value of a very slowly-converging infinite series...', *Phil. Trans.*, LXVIII part 2 (1778), pp. 895–901

Maseres, Francis (ed.) (1791–1807) *Scriptores Logarithmici;...*, 6 vols, London: vol. I (1791) *by* J. Davis; vol. II (1791) *by* J. Davis; vol. III (1796) *by* J. Davis, Wilks and Taylor; vol. IV (1801) *by* R. Wilks; vol. V (1804) *by* R. Wilks; vol. VI (1807) *by* R. Wilks

Maupertuis, Pierre-Louis Moreau de (1733) 'De figuris quas fluida rotata induere possunt, problemata duo; cum conjectura de stellis quae aliquando prodeunt vel deficiunt; & de Annulo Saturni', *Phil. Trans.*, XXXVII (1732), pp. 240–56

Millburn, John R. (1976) *Benjamin Martin; Author, Instrument-Maker, and 'Country Showman'*, Leyden: Noordhoff International Publishing

Milburn, John R. (1986) *Benjamin Martin; Author, Instrument-Maker, and 'Country Showman'*. Supplement, London: Vade-Mecum Press .

Miller, David P. (1983) 'Between hostile camps: Sir Humphry Davy's presidency of the Royal Society of London, 1820–1827', *The British Journal for the History of Science*, XVI, pp. 1–47

Miller, George (1799) *Elements of Natural Philosophy*, Dublin: *printed and sold by* R. E. Mercier

Milner, Isaac (1780) 'On the precession of the equinoxes produced by the Sun's attraction', *Phil. Trans.*, LXIX part 2 (1779), pp. 505–26

Mockler-Ferryman, A. K. (1900) *Annals of Sandhurst, a Chronicle of the Royal Military College from its Foundation to the Present Day with a Sketch of the History of the Staff College*, London: W. Heinemann

Moivre, Abraham de (1698a) 'Specimina quaedam illustria doctrinae fluxionum sive exempla quibus methodi istius usus & praestantia in solvendis problematis geometricis elucidatur', *Phil. Trans.*, XIX (1695), pp. 52–7

Moivre, Abraham de (1698b) 'A method of raising an infinite multinominal to any given power, or extracting any given root of the same', *Phil. Trans.*, XIX (1697), pp. 619–25

Moivre, Abraham de (1699) 'A method of extracting the root of an infinite equation', *Phil. Trans.*, XX (1698) pp. 190–3

Moivre, Abraham de (1704a) 'Methodus quadrandi genera quaedam curvarum, aut ad curvas simpliciores reducendi', *Phil. Trans.*, XXIII (1702), pp. 1113–27

Moivre, Abraham de (1704b) *Animadversiones in D. Georgii Cheynaei Tractatum de Fluxionum Methodo Inversa*, London: *typis* E. Midwinter, *impensis* D. Midwinter & T. Leigh

Moivre, Abraham de (1708) 'Aequationum quarumdam Potestatis tertiae, quintae, septimae, nonae, et superiorum, ad infinitum usque pergendo, in terminis finitis, ad instar regularum pro cubicis quae vocantur Cardani, resolutio analytica', *Phil. Trans.*, XXV (1707), pp. 2368–71

Moivre, Abraham de (1712) 'De mensura sortis, seu, de probabilitate eventuum in ludis a casu fortuito pendentibus', *Phil. Trans.*, XXVII (1711), pp. 206–64

Moivre, Abraham de (1717) 'A ready description and quadrature of a curve of the third order, resembling that commonly call'd the foliate', *Phil. Trans.*, XXIX (1715), pp. 329–31

Moivre, Abraham de (1718) *The Doctrine of Chances: or a Method of Calculating the Probability of Events in Play*, London: by W. Pearson, for the author

Moivre, Abraham de (1724a) 'De fractionibus algebraicis radicalitate immunibus ad fractiones simpliciores reducendis, deque summandis terminis quarumdam serierum aequali intervallo a se distantibus', *Phil. Trans.*, XXXII (1722), pp. 162–78

Moivre, Abraham de (1724b) 'De sectione anguli' *Phil. Trans.*, XXXII (1722), pp. 228–30

Moivre, Abraham de (1730) *Miscellanea Analytica de Seriebus et Quadraturis. Accessere Variae Considerationes de Methodis Comparationum, Combinationum & Differentiarum, Solutiones Difficiliorum aliquot Problematum ad Sortem Spectantium, itemque Construtiones faciles Orbium Planetarum, una cum Determinatione Maximarum & Minimarum Mutationum quae in Motibus Corporum Coelestium Occurrunt*, London: J. Tonson, J. Watts

Moivre, Abraham de (1741) 'De reductione radicalium ad simpliciores terminos, seu de extrahenda radice quacumque data ex Binomio $a + \sqrt{} + b$, vel $a + \sqrt{} - b$', *Phil. Trans.*, XL (1738), pp. 463–78

Montmort, Pierre Rémond de (1720) 'De seriebus infinitis tractatus. Pars prima', *Phil. Trans.*, XXX (1717), pp. 633–75

Muller, John (1736) *A Mathematical Treatise: Containing a System of Conic-Sections; with the Doctrine of Fluxions and Fluents, Applied to Various Subjects;...*, London: by T. Gardner for W. Innys, J. Nourse (French edn. 1760)

Muller, John (1746) *A Treatise Containing the Elementary Part of Fortification, Regular and Irregular [...] For the Use of the Royal Academy of Artillery at Woolwich*, London: J. Nourse

Muller, John (1747) *The Attack and Defence of Fortify'd Places. Containing, I. Preparations and Different Operations of an Attack. II. Preparations and Defence of Every Particular Part of a Fortification. III. A Treatise of Mines, Explaining the Manner of Making and Loading Them. For the Use of the Royal Academy of Artillery at Woolwich*, London: for J. Millan

Muller, John (1748) *Elements of Mathematics. Containing Geometry Conic-Sections Trigonometry Surveying Levelling Mensuration Laws of Motion Mechanics Projectiles Gunnery, &c. Hydrostatics Hydraulics Pneumatics A Theory of Pumps. To which is Prefix'd, The First Principles of Algebra, by Way of Introduction. For the Use of the Royal Academy of Artillery at Woolwich*, London: printed for the author and J. Millan

Muller, John (1755) *A Treatise Containing the Practical Part of Fortification. In Four Parts [...] For the Use of the Royal Academy of Artillery at Woolwich*, London: for A. Millar

Muller, John (1757) *A Treatise of Artillery: [...] to which is Prefixed, a Theory of Powder applied to Fire-Arms. For the Use of the Royal Academy of Artillery*, London:

for J. Millan (Appendix *Containing the True Projectile [...] the True Figure of the Earth,* 1768 (2nd edn. 1769; 3rd edn. 1780))

Murdoch, Patrick (1746) *Neutoni Genesis Curvarum per Umbras, seu Perspectivae Universalis Elementa; Exemplis Coni Sectionum et Linearum Tertii Ordinis Illustrata,* London: *apud* A. Millar

Murdoch, Patrick (1753) 'A letter from the Rev. Patrick Murdocke F.R.S. concerning the mean motion of the Moon's apogee, to the Rev. Dr. Robert Smith, Master of Trinity College, Cambridge', *Phil. Trans.,* XLVII (1751), pp. 62–74

Murdoch, Patrick (1769) 'An essay on the connexion between the parallaxes of the Sun and Moon; their densities; and their disturbing forces on the ocean', *Phil. Trans.,* LVIII (1768), pp. 24–33

Nagel, Ernst (1935) 'Impossible numbers: a chapter in the history of modern logic', *Studies in the History of Ideas,* III, pp. 429–74

Newton, Isaac (1687) *Philosophiae Naturalis Principia Mathematica,* London: Jussu Societatis Regiae *ac Typis* J. Streater (2nd edn. 1713; Amsterdam edn. 1714; 2nd Amsterdam edn. 1723; 3rd edn. 1726; 1st 'Le Seur, Jacquier' edn., Newton (1739, 40, 42); 2nd 'Le Seur, Jacquier' edn. 1760; 'Horsley' edn. in Newton (1779–85); 3rd 'Le Seur, Jacquier' edn. 1780–5; 1st English 'Motte' edn. 1729; 2nd English 'Thorpe' edn. Book I 1777; 1st French 'Madame du Chastellet' edn. 1756–9; 2nd French 'Madame du Chastellet' edn. 1759)

[Newton, Isaac] (1698) 'Epistola missa ad praenobilem virum D. Carolum Montague Armigerum, Scaccarii Regii apud Anglos Cancellarium, & Societatis Regiae Praesidem, in qua solvuntur duo problemata Mathematica à Johanne Bernoullo mathematico celeberrimo proposita', *Phil. Trans.,* XIX (1697), pp. 384–9

Newton, Isaac (1702) 'Lunae theoria Newtoniana', in D. Gregory (1702), pp. 332–6 (reprinted in Newton (1975))

Newton, Isaac (1704a) *Opticks: or, a Treatise of the Reflexions, Refractions, Inflexions and Colours of Light. Also two Treatises of the Species and Magnitude of Curvilinear Figures,* London: *for* S. Smith and B. Walford

Newton, Isaac (1704b) 'Enumeratio linearum tertii ordinis', in Newton (1704a), pp. 138–62 (reprinted in 1st Latin edn. Newton (1704a) 1706; English edn. in Harris (1704, 1710), II; another in Newton (1711a); another in Amsterdam edn. of Newton (1687) 1723; another in Newton (1744); another in Newton (1779–85))

Newton, Isaac (1704c) 'Tractatus de quadratura curvarum', in Newton (1704a), pp. 165–211 (reprinted in 1st Latin edn. Newton (1704a) 1706; 1st English edn. in Harris (1704, 1710), II; another in Newton (1711a); another in Amsterdam edn. Newton (1687) 1723; another in Newton (1744); 2nd English edn. in Newton (1745); another in Newton (1779–85))

Newton, Isaac (1711a) *Analysis per Quantitatum Series, Fluxiones, ac Differentias: cum Enumeratione Linearum Tertii Ordinis,* London: *ex officina* Pearsoniana (another in Amsterdam edn. Newton (1687) 1723; another? 1740)

Newton, Isaac (1711b) 'De analysi per aequationes numero terminorum infinitas', in Newton (1711a) (another in Collins (1713); another in Amsterdam edn.

Newton (1687) 1723; another in Newton (1744); English edn. in Newton (1745); another in Newton 1779–85))

Newton, Isaac (1711c) 'Methodus differentialis' in Newton (1711a) (another in Amsterdam edn. Newton (1687) 1723; another in Newton (1744); another in Newton (1779–85))

[Newton, Isaac] (1715) 'An account of the book entituled Commercium Epistolicum Collinii & aliorum, De Analysi promota; published by order of the Royal Society, in relation to the dispute between Mr. Leibnitz and Dr. Keill, about the right of invention of the method of fluxions, by some call'd the differential method', Phil. Trans., XXIX (1715), pp. 173–224

[Newton, Isaac] (1717) 'Problematis mathematicis anglis nuper propositi solutio generalis', Phil. Trans., XXIX (1716), pp. 399–400

Newton, Isaac (1736) The Method of Fluxions and Infinite Series; with its Application to the Geometry of Curve-Lines. By the Inventor Sir Isaac Newton, K'. Late President of the Royal Society. Translated from the Author's Latin Original not yet made publick. To which is subjoin'd, A Perpetual Comment upon the whole Work, Consisting of Annotations, Illustrations, and Supplements, In order to make this Treatise a Compleat Institution for the use of Learners (translation and notes John Colson), London: by H. Woodfall, sold J. Nourse (pirated edn. Newton (1737); French transl. by Buffon 1740; Latin edn. in Newton (1744); another Latin edn. in Newton (1779–85))

Newton, Isaac (1737) A Treatise of the Method of Fluxions and Infinite Series, with its Application to the Geometry of Curve Lines. By Sir Isaac Newton, Kt. Translated from the Latin Original not yet published. Designed by the Author for the Use of Learners, London: for T. Woodward and J. Millan

Newton, Isaac (1739, 1740, 1742) Philosophiae Naturalis Principia Mathematica auctore Isaaco Newtono, Eq. Aurato. Perpetuis Commentariis illustrata, communi studio PP. Thomae Le Seur & Francisci Jacquier Ex Gallicana Minimorum Familia, Matheseos Professorum, 3 vols, Genéve: Typis Barrillot & Filii

Newton, Isaac (1744) Opuscola Mathematica, Philosophica et Philologica (ed. Castiglione), Lausanne and Genéve: Bousquet & Socios

Newton, Isaac (1745) Sir Isaac Newton's Two Treatises of the Quadrature of Curves, and Analysis by Equations of an Infinite Number of Terms, Explained: Containing The Treatises themselves, Translated into English, with a Large Commentary; in which the Demonstrations are Supplied where Wanting, the Doctrine Illustrated, and the whole Accommodated to the Capacities of Beginners, for whom it is chiefly designed (trans. and ed. John Stewart), London: by J. Bettenham, for Society for the Encouragement of Learning, sold by J. Nourse and John Whiston, London

Newton, Isaac (1779–85) Isaaci Newtoni Opera Quae Exstant Omnia... (ed. Samuel Horsley), 5 vols, London: excudebant J. Nichols

Newton, Isaac (1959–77) The Correspondence of Isaac Newton (eds. H. W. Turnbull et al.), 7 vols, Cambridge University Press

Newton, Isaac (1967–81) The Mathematical Papers of Isaac Newton (ed. D. T. Whiteside), 8 vols, Cambridge University Press

Newton, Isaac (1975) Isaac Newton's Theory of the Moon's Motion (1702) (ed. I. B. Cohen), Folkestone: Dawson

Newton, Thomas (1805) An Illustration of Sir Isaac Newton's Method of Reasoning by

Prime and Ultimate Ratios; Comprehending the First Section of his Principia, and so much of the Second and Third Sections as is Necessary to Explain the Motions of the Heavenly Bodies, Leeds: by E. Baines

Olson, Richard (1971) 'Scottish philosophy and mathematics: 1750–1830', *Journal for the History of Ideas*, XXXII, pp. 29–44

Ozanam, Jacques (1691) *Dictionnaire Mathématique; ou Idée Generale des Mathématiques*, Paris: E. Michallet

Paman, Roger (1745) *The Harmony of the Ancient and Modern Geometry Asserted: In Answer to the Call of the Author of the Analyst upon the Celebrated Mathematicians of the Present Age, to clear up what he Stiles, their Obscure Analytics*, London: J. Nourse

Panteki, Maria (1987) 'William Wallace and the introduction of continental calculus to Britain: a letter to George Peacock', *Historia Mathematica*, XIV, pp. 119–32

Panza, Marco (1989) *La Statua di Fidia, Analisi Filosofica di una Teoria Matematica: il Calcolo delle Flussioni*, Milano: Unicopli

Parkinson, Thomas (1785) *A System of Mechanics, Being the Substance of Lectures upon that Branch of Natural Philosophy*, Cambridge: J. Archdeacon

Peacock, George (1834) 'Report on the recent progress and present state of certain branches of analysis', *Report of the Third Meeting of the British Association for the Advancement of Science*, pp. 185–352, London: J. Murray

Pedersen, Olaf (1963) 'The "philomaths" of 18th century England', *Centaurus*, VIII, pp. 238–62

Pemberton, Henry (1722) *Epistola ad Amicum de Cotesii Inventis, Curvarum Ratione, quae cum Circulo & Hyperbola Comparationem Admittunt*, London: impensis G. & J. Innys (Appendix 1723)

[Pemberton, Henry] (1724) 'Solutio problematis de curvis inveniendis, quae quadam ratione in situ inverso dispositae se intersecare possunt in angulo dato', *Phil. Trans.*, XXXII (1722), pp. 106–38

Pemberton, Henry (1728) *A View of Sir Isaac Newton's Philosophy*, London: by S. Palmer

Pemberton, Henry (1772) 'Geometrical solutions of three celebrated astronomical problems', *Phil. Trans.*, LXII (1772), pp. 434–46

Perks, John (1708) 'The construction and properties of a new quadrature to the hyperbola', *Phil. Trans.*, XXV (1706), pp. 2253–62

Perks, John (1717) 'An easy mechanical way to divide the nautical meridian line in Mercator's projection; with an account of the relation of the same meridian line to the curva catenaria', *Phil. Trans.*, XXIX (1715), pp. 331–9

Petvin, John (1750) *Letters concerning Mind. To Which is Added, a Sketch of Universal Arithmetic; Comprehending the Differential Calculus, and the Doctrine of Fluxions*, London: for J. and J. Rivington

Playfair, John (1779) 'On the arithmetic of impossible quantities', *Phil. Trans.*, LXVIII part 1 (1778), pp. 318–43

Playfair, John (1788a) 'Account of Matthew Stewart DD', *Edin. Trans.*, I part 1 (1788), pp. 57–76 (reprinted in Playfair (1822a)), IV, pp. 2–30

Playfair, John (1788b) 'On the causes which affect the accuracy of barometrical measurements', *Edin. Trans.*, I part 2 (1788), pp. 87–130

Playfair, John (1795) *Elements of Geometry; Containing the First Six Books of Euclid, With Two Books on the Geometry of Solids. To Which are Added, Elements of Plane and Spherical Trigonometry*, Edinburgh: *for* Bell and Bradfute, G. G. & J. Robinson

Playfair, John (1805) 'Investigation of certain theorems relating to the figure of the Earth', *Edin. Trans.*, V part 1 (1805), pp. 3–30

Playfair, John (1808) 'Traité de Mechanique Celeste', *The Edinburgh Review*, no. XXII, pp. 249–84

Playfair, John (1812) 'Of the solids of greatest attraction, or those which, among all the solids that have certain properties, attract with the greatest force in a given direction', *Edin. Trans.*, VI part 1 (1812), pp. 187–247

Playfair, John (1812–14) *Outlines of Natural Philosophy Being Heads of Lectures Delivered in the University of Edinburgh*, 2 vols, Edinburgh: *by* A. Neill &c.

[Playfair, John] (1816) 'The principles of fluxions, designed for the use of the students in the universities, by William Dealtry', *The Edinburgh Review*, pp. 87–98

Playfair, John (1822a) *The Works of John Playfair*, 4 vols, Edinburgh: A. Constable and Hurst

Playfair, John (1822b) 'A general view of the progress of mathematical and physical science, since the revival of letters in Europe', in Playfair (1822a), II, pp. 5–445 (also published in vol. II and vol. IV of *Encyclopaedia Britannica* supplementary volumes to the 4th, 5th and 6th edns. (1824))

Poisson, Siméon D. (1811) *Traité de Méchanique*, Paris: Courcier

Ponting, Betty (1979a) 'Mathematics at Aberdeen, developments, characters and events, 1495–1717', *Aberdeen University Review*, no. 161, pp. 26–35

Ponting, Betty (1979b) 'Mathematics at Aberdeen, developments, characters and events, 1717–1860', *Aberdeen University Review*, no. 162, pp. 162–76

Porter, Withworth, and Watson, Charles (1889–1915) *History of the Corps of Royal Engineers*, 3 vols, London and New York

Rait, Robert Sangster (1895) *The University of Aberdeen, a History*, Aberdeen: *by* J. G. Bisset

Raphson, Joseph (1697) *Analysis Aequationum Universalis seu Ad Aequationes Algebraicas Resolvendas Methodus Generalis, & Expedita, Ex nova Infinitorum Serierum Methodo, Deducta ac Demonstrata. Editio Secunda cum Appendice. Cui Annexum est, de Spatio Reali, seu Ente Infinito Conamen Mathematico-Metaphysicum*, London: T. Braddyll

Raphson, Joseph (1702) *A Mathematical Dictionary: or, a Compendious Explication of all Mathematical Terms, Abridg'd from Monsieur Ozanam, and others. With a Translation of his Preface, and an Addition on Several Easie and Useful Abstracts; on Plain Trigonometry, Mechanicks, the first Properties of the three Conick Sections, &c. To which is added, an Appendix, containing the Quantities of all Sorts of Weights and Measures, the Explanation of the Characters used in Algebra. Also the Definition and Use of the Principal Mathematical Instruments, and the Instruments Themselves Curiously Engraved in Copper*, London: J. Nicholson, T. Leigh, D. Midwinter

Raphson, Joseph (1715) *The History of Fluxions, Shewing in a Compendious Manner the first Rise of, and various Improvements made in that Incomparable Method*, London: *by* W. Pearson, *sold by* R. Mount (1715 Latin edn.; 2nd edn. 1717)

Rigaud, Stephen (ed.) (1841) *Correspondence of Scientific Men of the Seventeenth Century...*, 2 vols, Oxford University Press

Rivers, Isabelle (ed.) (1982) *Books and Their Readers in Eighteenth-Century England*, Leicester University Press

Robartes, Francis (1712) 'Concerning the proportion of mathematical points to each other', *Phil. Trans.*, XXVII (1712), pp. 470–2

Robertson, Abram (1795) 'The binomial theorem demonstrated by the principles of multiplication', *Phil. Trans.*, LXXXV, part 2 (1795), pp. 298–321

Robertson, Abram (1806) 'A new demonstration of the binomial theorem, when the exponent is a positive or negative fraction', *Phil. Trans.*, XCVI, part 2 (1806), pp 305–26

Robertson, Abram (1807) 'On the precession of the equinoxes', *Phil. Trans*, XCVII, part 1 (1807), pp. 57–82

Robertson, John (1754) *The Elements of Navigation; Containing the Theory and the Practice. With all the Necessary Tables. To which is added, a Treatise of Marine Fortification. For the Use of the Royal Mathematical School at Christ's Hospital, and the Gentlemen of the Navy*, London: for J. Nourse

Robins, Benjamin (1728) 'A demonstration of the 11th proposition of Sir Isaac Newton's Treatise of Quadratures', *Phil. Trans.*, XXXIV (1827), pp. 230–6

Robins, Benjamin (1735) *A Discourse Concerning the Nature and Certainty of Sir Isaac Newton's Methods of Fluxions, and of Prime and Ultimate Ratios*, London: for W. Innys and R. Manby (reprinted in Robins (1761))

Robins, Benjamin (1742) *New Principles of Gunnery: Containing the Determination of the Force of Gun-Powder, and an Investigation of the Difference in the Resisting Power of the Air to Swift and Slow Motions*, London: for J. Nourse (German edn. enlarged Robins (1745))

[Robins, Benjamin] (1744) 'An account of a book intituled, New Principles of Gunnery, Containing the Determination of the Force of Gunpowder, and an Investigation of the Resisting Power of the Air to Swift and Slow Motions; by B.R. F.R.S. as far as the same relates to the force of gunpowder', *Phil. Trans.*, XLII (1743), pp. 437–56

Robins, Benjamin (1745) *Neue Grundsätze der Artillerie*, Berlin: A Haude (translation of Robins (1742) extensively annotated by Euler; 1st French edn. 1771; 2nd French edn. 1783; Euler's notes translated into English 1777)

Robins, Benjamin (1761) *Mathematical Tracts of the late Benjamin Robins, Esq. Fellow of the Royal Society, and Engineer General to the Honourable the East India Company. In two Volumes. Vol. I Containing his New Principles of Gunnery, with Several Subsequent Discourses on the same Subject, the Greatest Part never before Printed. Vol. II Containing his Discourse on the Methods of Fluxions, and of Prime and Ultimate Ratios, with other Miscellaneous Pieces* (ed. James Wilson), London: J. Nourse

Robinson, Thomas R. (1820) *A System of Mechanics, for the use of the Students in the University of Dublin*, Dublin University Press, for Hodges and M'Arthur

Robison, John (1804) *Elements of Mechanical Philosophy, Being the Substance of a Course of Lectures on that Science, Volume First, Including Dynamics and Astronomy*, Edinburgh: for A. Constable, *et al.*

Rousseau, G.S. (1982) 'Science books and their readers in the eighteenth century', in Rivers (1982)

Rowe, John (1751) *An Introduction to the Doctrine of Fluxions*, London: *by* E. Owen, *sold by* J. Noon (2nd edn. 1757; another? 1762; 3rd edn. 1767; 4th edn. 1809)

Rowning, John (1744, 1745) *A Compendious System of Natural Philosophy, with Notes Containing the Mathematical Demonstrations and some Occasional Remarks, in four parts*, London: *for* S. Harding (already published in parts by Cambridge University Press 1734–)

Rowning, John (1756) *A Preliminary Discourse to an intended Treatise on the Fluxionary Method. Largely Explaining the Nature and Peculiarity of that Doctrine, In a Familiar and Easy Manner*, London: *for* S. Harding, *sold by* B. Dod and J. Marks

Rutherforth, Thomas (1748) *A System of Natural Philosophy, being a Course of Lectures in Mechanics, Optics, Hydrostatics, and Astronomy; Which are read in St. John's College Cambridge*, Cambridge: J. Bentham

Salmon, Thomas (1749) *A New Geographical and Historical Grammar: Wherein the Geographical Part is Truly Modern; and the Present State of the Several Kingdoms of the World is so Interspersed, as to Render the Study of Geography both Entertaining and Instructive...*, London: *for* W. Johnston

Sandhurst (1802) *His Majesty's Warrants: Containing the Appointment and Supreme Board of Commissioners for the Affairs of the Royal Military College; and Statutes for the Government and Conduct of the First Department of that Institution*, London: *by* March

Sandhurst (1809) *Warrant for Regulating All Matters Relative to the Royal Military College*, London: *by* T. Egerton

Sandhurst (1849) *Complete Guide to the Junior and Senior Departments of the Royal Military College, Sandhurst*, London: *by* C. Law

Sault, Richard (1699) 'Curvae celerrimi descensus investigatio analytica excerpta ex literis R. Sault, Math.P.', *Phil. Trans.*, XX (1698), pp. 425–6

Saunderson, Nicholas (1740) *The Elements of Algebra, in Ten Books, by Nicholas Saunderson LLD [...], to which is prefixed I. The Life and Character of the Author. II. His Palpable Arithmetic Decyphered*, Cambridge University Press

Saunderson, Nicholas (1756a) *The Method of Fluxions Applied to a Select Number of Useful Problems: Together with the Demonstration of Mr. Cotes's Forms of Fluents in the Second Book of his Logometria; The Analysis of the Problems in his Scholium Generale; and an Explanation of the principal Propositions of Sir Isaac Newton's Philosophy*, London: *for* A. Millar, J. Whiston and B. White, L. Davies and C. Reymers

Saunderson, Nicholas (1756b) *Select Parts of Professor Saunderson's Elements of Algebra. For the use of Students at the Universities*, London: *by* J. Kippax, *for* A. Millar, *et al.*

Schneider, I. (1968) 'Der Mathematiker Abraham De Moivre', *Archive for History of Exact Sciences*, V, pp. 177–317

Scott, George P. (1971) *Some Aspects of the Mathematical Work of Colin MacLaurin (1698–1746)*, M.Sci. thesis, London University

Scriba, Christoph J. (1964) 'The inverse method of tangents. A dialogue between Leibniz and Newton (1675–1677)', *Archive for History of Exact Sciences*, II, pp. 113–37

Scriba, Christoph J. (1969) 'Neue Dokumente zur Entstehungsgeschichte des Prioritätsstreites zwischen Leibniz und Newton um die Erfindung der Infinitesimalrechnung', *Akten des Internationalen Leibniz-Kongresses Hannover, 14–19 November 1966*, II, pp. 69–78

Sewell, William (1796) 'Newton's binomial theorem legally demonstrated by algebra', *Phil. Trans.*, LXXXVI part 2 (1796), pp. 382–4

Short, J. (1754) 'An account of a book intitled P.D. Pauli Frisii Mediolanensis, &c. Disquisitio Mathematica in Causam Physicam Figurae et Magnitudinis Telluris Nostrae; printed at Milan 1752. Inscribed to the Count de Sylva, and consisting of ten sheets and a half in quarto, *Phil. Trans.*, XLVIII part 1 (1753), pp. 5–17

Silvabelle, de St. Jacques (1755) 'A treatise on the precession of the equinoxes, and in general on the motion of the nodes, and the alteration of the inclination of the orbit of a planet to the ecliptic', *Phil. Trans.*, XLVIII part 2 (1754), pp. 385–441 (trans. J. Bevis)

Simpson, Thomas (1737) *A New Treatise of Fluxions: wherein the Direct and Inverse Method are Demonstrated after a New, Clear, and Concise Manner, With their Application to Physics and Astronomy: also the Doctrine of Infinite Series and Reverting Series Universally, are Amply Explained, Fluxionary and Exponential Equations Solved: Together with a Variety of New and Curious Problems*, London: by T. Gardner, for the author, G. Powell, R. Shirtcliffe, D. England

Simpson, Thomas (1740) *Essays on Several Curious and Useful Subjects, in Speculative and Mix'd Mathematicks, Illustrated by a Variety of Examples*, London: by H. Woodfall, for J. Nourse

Simpson, Thomas (1743) *Mathematical Dissertations on a Variety of Physical and Analytical Subjects...*, London: for T. Woodward

Simpson, Thomas (1750a) 'The motion of projectiles near the Earth's surface consider'd, independent of the properties of the conic sections', *Phil. Trans.*, XLV (1748), pp. 137–47

Simpson, Thomas (1750b) 'Of the fluents of multinomials, and series affected by radical signs, which do not begin to converge till after the second term', *Phil. Trans.*, XLV (1748), pp. 328–35

Simpson, Thomas (1750c) *The Doctrine and Application of Fluxions, Containing (Beside what is Common on the Subject) a Number of New Improvements in the Theory. And the Solution of a Variety of New, and very Interesting, Problems in different Branches of the Mathematicks*, 2 vols, London: for J. Nourse (2nd edn. enlarged 1776; 3rd edn. 1805; 4th edn. revised and adapted 1823)

Simpson, Thomas (1752) *Select Exercises for Young Proficients in the Mathematicks. Containing...*, London: for J. Nourse (2nd edn. 1792, 3rd edn. 1810)

Simpson, Thomas (1753) 'A general method for exhibiting the value of an algebraic expression involving several radical quantities in an infinite series: wherein Sir Isaac Newton's theorem for involving a binomial, with another of the same author, relating to the roots of equations, are demonstrated', *Phil. Trans.*, XLVII (1751), pp. 20–7

Simpson, Thomas (1756a) 'An investigation of a general rule for the resolution of isoperimetrical problems of all orders', *Phil. Trans.*, XLIX part 1 (1755), pp. 4–15

Simpson, Thomas (1756b) 'A letter to the Right Honourable George Earl of Macclesfield, President of the Royal Society, on the advantage of taking the mean of a number of observations, in practical astronomy, *Phil. Trans.*, XLIX part 1 (1755), pp. 82–93

Simpson, Thomas (1757) *Miscellaneous Tracts on Some Curious, and Very interesting Subjects in Mechanics, Physical-Astronomy, and Speculative Mathematics; wherein, the Precession of the Equinox, the Nutation of the Earth's Axis, and the Motion of the Moon in her Orbit, are Determined,* London: *for* J. Nourse

Simpson, Thomas (1758) 'The resolution of a general proposition for determining the horary alteration of the position of the terrestrial equator, from the attraction of the Sun and Moon: with some remarks on the solutions given by other authors to that difficult and important problem', *Phil. Trans.*, L part 1 (1757), pp. 416–27

Simpson, Thomas (1759a) 'A further attempt to facilitate the resolution of isoperimetrical problems', *Phil. Trans.*, L part 2 (1758), pp. 623–31

Simpson, Thomas (1759b) 'The invention of a general method for determining the sum of every 2d, 3d, 4th, or 5th, &c. term of a series, taken in order; the sum of the whole series being known', *Phil. Trans.*, L part 2 (1758), pp. 757–69

Simpson, Thomas (1760) *Elements of Geometry, with their Application to the Mensuration of Superficies and Solids, to the Determination of the Maxima and Minima of Geometrical Quantities, and to the Construction of a Great Variety of Geometrical Problems,* 2nd edn., London: *for* J. Nourse

Simson, Robert (1724) 'Pappi Alexandrini propositiones duae generales, quibus plura ex Euclidis porismatis complexus est', *Phil. Trans.*, XXXII (1723), pp. 330–40

Simson, Robert (1776) *Opera Quaedam Reliqua, scilicet, I. Apollonii Pergaei de sectione determinata libri II. restituti, duobus insuper libris aucti. II. Porismatum liber, quo doctrinam hanc veterum geometrarum ad oblivione vindicare, et ad captum hodiernorum adumbrare constitutum est. III. de logarithmis liber. IV. De Limitibus Quantitatum et Rationum, fragmentum. V. Appendix pauca continens problemata ad illustrandam praecipue veterum geometrarum analysin. nunc primum post auctoris mortem in lucem edita...*, Glasgow: R. & A. Foulis (part IV *de limitibus...* reprinted in Maseres (ed.) (1791–1807), vol. VI (1807))

Smeaton, John (1769) 'A discourse concerning the menstrual parallax, arising from the mutual gravitation of the Earth and Moon; its influence on the observations of the Sun and planets; with a method of observing it', *Phil. Trans.*, LVIII (1768), pp. 156–69

Smith, G. C. (1980) 'Thomas Bayes and Fluxions', *Historia Mathematica*, VII, pp. 379–88

Smith, James (1737) *A New Treatise of Fluxions. Containing, I the Elements of Fluxions, demonstrated in Two easy Propositions, without first or last Ratios. II A Treatise of Nascent and Evanescent Quantitates, first and last Ratios. III Sir Isaac Newton's Demonstrations of the Fluxions enlarged and illustrated. IV Answers to the Principal Objections in the Analyst,* London: *printed for* the author

Smith, Robert (1738) *A Compleat System of Opticks. In Four Books, viz. A Popular, A Mathematical, A Mechanical, and A Philosophical Treatise. To which are Added Remarks upon the Whole,* Cambridge: *for* the author

Smyth, John (1961) *Sandhurst – The History of the Royal Military Academy,*

*Woolwich, The Royal Military College, Sandhurst, and the Royal Military Academy,
Sandhurst 1741–1961*, London: Weidenfeld & Nicolson

Spence, William (1809) *An Essay on the Theory of the Various Orders of Logarithmic
Transcendents; with an Inquiry into their Applications to the Integral Calculus and
the Summation of Series*, London: *for* J. Murray and A. Constable

Spence, William (1814) *Outlines of a Theory of Algebraical Equations, Deduced from
the Principles of Harriott, and Extended to the Fluxional or Differential Calculus*,
London: *for* the author, *by* Davis and Dickson

Spence, William (1820) *Mathematical Essays, by the Late William Spence, Esq. Edited
by John F.W. Herschel, Esq. with a Brief Memoir of the Author*, London: *by*
J. Moyes *for* Oliver & Boyd, G. & W.B. Whittaker (1st edn. 1919?)

Stack, J. (1793) *A Short System of Optics*, Dublin University Press

Stewart, Matthew (1746) *Some General Theorems of Considerable Use in the Higher
Parts of Mathematics*, Edinburgh: *for* W. Sands, A. Murray, J. Cochran

Stewart, Matthew (1754) 'Pappi Alexandrini collectionum mathematicarum libri
quarti propositio quarta generalior facta, cui propositiones aliquot eodem
spectantes adjicuntur', *Essays and Observations Physical and Literary, Read before
a Society in Edinburgh, and published by them*, I, pp. 141–72

Stewart, Matthew (1756) 'A solution of Kepler's problem', *Essays and Observations
Physical and Literary, Read before a Society in Edinburgh, and published by them*, II,
pp. 105–44

Stewart, Matthew (1761) *Tracts, Physical and Mathematical. Containing, an
Explication of Several Important Points in Physical Astronomy; and, a New Method
for Ascertaining the Sun's Distance from the Earth, by the Theory of Gravity*,
Edinburgh: *for* A. Millar, J. Nourse, W. Sands, A. Kincaid, J. Bell

Stewart, Matthew (1763a) *Propositiones Geometricae, More Veterum demonstratae,
Ad Geometriam Antiquam Illustrandam et Promovendam Idoneae*, Edinburgh: *typis*
Sands, Murray, Cochran

Stewart, Matthew (1763b) *The Distance of the Sun from the Earth Determined by the
Theory of Gravity. Together with Several Other Things Relative to the Same Subject,
by Dr. Matthew Stewart, Professor of Mathematics in the University of Edinburgh.
Being a Supplement to Tracts Physical and Mathematical, lately published by the same
Author*, Edinburgh: *for* A. Millar, J. Nourse, D. Wilson, W. Sands, A. Kincaid and
J. Bell

Stirling, James (1717) *Lineae Tertii Ordinis Neutonianae, sive Illustratio Tractatus D.
Neutoni De Enumeratione Linearum Tertii Ordinis, Cui Subjiungitur, Solutio Trium
Problematum*, Oxford: *e* Theatro Sheldoniano, *impensis* E. Whisler

Stirling, James (1720) 'Methodus Differentialis Newtoniana Illustrata', *Phil Trans.*,
XXX (1719), pp. 1050–70

Stirling, James (1730) *Methodus Differentialis: sive Tractatus de Summatione et
Interpolatione Serierum Infinitarum*, London: *typis* Bowyer, *impensis* G. Strahan
(2nd edn. 1753; 3rd edn. 1764; English translation by Francis Holliday 1749)

Stirling, James (1738) 'Of the figure of the Earth, and the variation of gravity on
the surface', *Phil. Trans.*, XXXIX (1735), pp. 98–105

Stone, Edmund (1726) *A New Mathematical Dictionary: Wherein is contain'd, not
only the Explanation of the Bare Terms, But likewise an History, of the Rise, Progress,*

State, Properties, &c. of Things, both in Pure Mathematicks, and Natural Philosophy,
so far as it comes Under a Mathematical Consideration, London: J. Senex, W. &
J. Innys, J. Osborn, T. Longman, T. Woodward (2nd edn. 1743)

Stone, Edmund (1730) *The Method of Fluxions both Direct and Inverse. The Former*
being a Translation from the Celebrated Marquis De L'Hospital's Analyse des
Infinements Petits: and the Latter Supply'd by the Translator, 2 vols, London: *for*
W. Innys

Stone, Edmund (1744) 'A letter from Edmund Stone, F.R.S. to ———— concerning
two species of lines of the third order, not mentioned by Sir Isaac Newton, nor
Mr. Stirling', *Phil. Trans.,* XLI part 1 (1740), pp. 318–20

Taylor, Brook (1714a) 'De inventione centri oscillationis', *Phil. Trans.,* XXVIII
(1713), pp. 11–21

Taylor, Brook (1714b) 'De motu nervi tensi', *Phil. Trans.,* XXVIII (1713),
pp. 26–32

Taylor, Brook (1715) *Methodus Incrementorum Directa & Inversa,* London: *typis*
Pearsoniani, *for* W. Innys (2nd edn. 1717)

Taylor, Brook (1717) 'An account of a book entituled Methodus Incrementorum, auctore Brook Taylor LLD. & R.S. Secr.', *Phil. Trans.,* XXIX (1715),
pp. 339–50

Taylor, Brook (1720a) 'An attempt towards the improvement of a method of
approximating, in the extraction of the roots of equations in numbers', *Phil.*
Trans., XXX (1717), pp. 610–22

Taylor, Brook (1720b) 'Solutio problematis a Dom^{no} G.G. Leibnitio geometris
anglis nuper propositi', *Phil. Trans.,* XXX (1717), pp. 695–701

Taylor, Brook (1720c) 'Appendix. Qua methodo diversa eadem materia tractatur',
Phil. Trans., XXX (1717), pp. 676–89

Taylor, Brook (1723) 'Propositiones aliquot de projectilium motu parabolico,
scriptae an. 1710', *Phil. Trans.,* XXXI (1721), pp. 151–63

Taylor, E.G.R. (1954) *The Mathematical Practitioners of Tudor and Stuart England,*
Cambridge University Press, *for the* Institute of Navigation

Taylor, E.G.R. (1966) *Mathematical Practitioners of Hanoverian England*
1714–1840, Cambridge University Press, *for the* Institute of Navigation

Taylor, W.B.S. (1845) *History of the University of Dublin,* London: *for* T. Caddel

Thomas, Hugh S. (1961) *The Story of Sandhurst,* London: Hutchinson

Todhunter, Isaac (1873) *A History of the Mathematical Theories of Attraction and the*
Figure of the Earth, 2 vols, London: Macmillan & co.

Toplis, John (1805) 'On the decline of mathematical studies, and the sciences
dependent upon them', *Philosophical Magazine,* XX, pp. 25–31

Townshend, George (1776) *Rules and Orders for the Royal Military Academy at*
Woolwich

Trail, William (1796) *Elements of Algebra, For the Use of Students in Universities…,*
4th edn., Edinburgh: *for* W. Creech, *sold by* Robinson, T. Kay

Trail, William (1812) *Account of the Life and Writings of Robert Simson, MD, late*
Professor of Mathematics in the University of Glasgow, Bath: *by* R. Cruttwell *for*
G. & W. Nicol

Truesdell, Clifford A. (1954) 'Rational fluid mechanics, 1687–1765', in L. Euler
Opera Omnia, Lausanne, ser 2, XII, pp. ix–cxxv

Turnbull, Herbert W. (1951) 'Bi-centenary of the death of Colin Maclaurin (1698–1746)', *Aberdeen University Studies*, no. 127, Aberdeen University Press

Turner, G.L'E. (1986) 'The physical sciences', in *The History of the University of Oxford*, V, pp. 659–81, Oxford: Clarendon Press

Tweedie, Charles (1915) 'A study of the life and writings of Colin Maclaurin', *The Mathematical Gazette*, VIII, pp. 133–51

Tweedie, Charles (1922) *James Stirling: A Sketch of His Life and Works Along with His Scientific Correspondence*, Oxford: Clarendon Press

Vauban, Sébastien Le Prestre de (1737, 42) *De L'attaque et de la Defense des Places*, 2 vols, La Haye: P. de Hondt

Venturoli, Giuseppe (1822) *Elements of the Theory of Mechanics...*, (trans. Daniel Cresswell), Cambridge: by J. Smith, for J. Nicholson & son

Venturoli, Giuseppe (1823) *Elements of Practical Mechanics [...] To which is added a Treatise upon the Principle of Virtual Velocities, and its Uses in Mechanics*, (trans. Daniel Cresswell), Cambridge: by J. Smith, for J. Deighton

Vince, Samuel (1781) 'An investigation of the principles of progressive and rotatory motion', *Phil. Trans.*, LXX part 2 (1780), pp. 546–77

Vince, Samuel (1783) 'A new method of investigating the sums of infinite series', *Phil. Trans.*, LXXII part 2 (1782), pp. 389–416

Vince, Samuel (1785) 'A supplement to the third part of the paper on the summation of infinite series, in the Philosophical Transactions for the year 1782', *Phil. Trans.*, LXXV part 1 (1785), pp. 32–5

Vince, Samuel (1786) 'A new method of finding fluents by continuation', *Phil. Trans.*, LXXVI part 2 (1786), pp. 432–42

Vince, Samuel (1787) 'On the precession of equinoxes', *Phil. Trans.*, LXXVII part 2 (1787), pp. 363–7

Vince, Samuel (1790) *A Treatise on Practical Astronomy*, Cambridge: by J. Archdeacon *for* J. & J. Merrill

Vince, Samuel (1795) *The Principles of Fluxions: Designed for the Use of Students in the University*, in Vince and Wood (1795–9), II (1st American edn. 1798; 2nd edn. 1800; 3rd edn. corrected and enlarged 1805; 4th edn. 1812; 2nd American edn. 1812; 5th edn. 1818)

Vince, Samuel (1797, 1799, 1808) *A Complete System of Astronomy*, 3 vols, Cambridge University Press

Vince, Samuel, and Wood, James (1795–9) *The Principles of Mathematics and Natural Philosophy*, 4 vols, Cambridge: J. Burges

Waff, Craig B. (1976) *Universal Gravitation and the Motion of the Moon's Apogee: The Establishment and Reception of Newton's Inverse-Square Law, 1687–1749*, Ph.D., The Johns Hopkins University, Baltimore

Wallace, William (1798) 'Some geometrical porisms, with examples of their application to the solution of problems', *Edin. Trans.*, IV part 2 (1798), pp. 107–34

Wallace, William (1805) 'A new method of expressing the coefficients of the development of the algebraic formula $(a^2 + b^2 - 2ab\cos\phi)^n$, by means of the perimeters of two Ellipses, when n denotes the half of any odd number; together with an appendix, containing the investigation of a formula for the rectification of any arch of an ellipse', *Edin. Trans.*, V part 2, (1805), pp. 253–93

Wallace, William (1810) 'Fluxions', *Encyclopaedia Britannica*, 4th edn., VIII, pp. 697–778

Wallace, William (1815) 'Fluxions', *Edinburgh Encyclopaedia*, IX, pp. 382–467

Wallis, John (1693) *Opera Mathematica*, II, Oxford: *e* Theatro Sheldoniano

Wallis, John (1699) *Opera Mathematica*, III, Oxford: *e* Theatro Sheldoniano

Wallis, Peter J. (1973) 'British philomaths – mid-eighteenth-century and earlier', *Centaurus*, XVII, pp. 301–14

Wallis, Peter J. (1976) *An Index of British Mathematicians. Part 2. 1701–60*, PHIBB 105, University of Newcastle upon Tyne

Wallis, Peter J. (1982) 'The MacLaurin "circle" – the evidence of subscription lists', *The Bibliotheck*, XI, pp. 38–54

Wallis, Peter J., and Wallis, Ruth V. (1986) *Biobibliography of British Mathematics and its Application, Part II, 1701–1760*, PHIBB, Project for Historical Bio-bibliography, Newcastle upon Tyne: Epsilon Press

Walmesley, Charles (1749a) *Théorie du Mouvement des Apsides en générale, et en particulier des Apsides de l'Orbite de la Lune*, Paris: G.F. Quillau

Walmesley, Charles (1749b) *Analyse des Mesures des rapports et des angles: ou, Reduction des Integrales aux Logarithmes, et aux Arcs de Cercle*, Paris: G.F. Quillau (another 1753)

Walmesley, Charles (1754) *The Theory of the Motion of the Apsides in General, and of the Apsides of the Moon's Orbit in Particular*, London: *for* W. Owen (translation of Walmesley (1749a) by J. Brown and revised by William Emerson)

Walmesley, Charles (1757) 'Two essays addressed to the Rev. James Bradley DD and Astrom. Reg. by Mr. Charles Walmesley, FRS', *Phil. Trans.*, XLIX part 2 (1756), pp. 700–58 (The two essays are 'De Praecessione Aequinotiorum et axis Terrae Nutatione', pp. 704–36, and 'De Inaequalitatibus motuum Terrae', pp. 737–58)

Walmesley, Charles (1758) *De Inaequalitatibus Motuum Lunarium*, Firenze: *e* typographio imperiali

Walmesley, Charles (1759) 'Of the irregularities in the motion of a satellite arising from the spheroidical figure of its primary planet', *Phil. Trans.*, L part 2 (1758), pp. 809–35

Walmesley, Charles (1762) 'Of the irregularities in the planetary motions, caused by the mutual attraction of the planets', *Phil. Trans.*, LII part 1 (1761), pp. 275–335

Walter, Thomas (1762?) *A New mathematical Dictionary …*, London: *for and sold by* the author

Walton, Jakob (1735a) *A Vindication of Sir Isaac Newton's Principles of Fluxions, against the Objections contained in the Analyst*, Dublin: *by* J. Powell *for* W. Smith

Walton, Jakob (1735b) *The Catechism of the Author of the Minute Philosopher Fully Answer'd*, Dublin: *by* S. Powell, *for* W. Smith (2nd edn. *With an Appendix in Answer to the Reasons for not Replying to Mr. Walton's Full Answer, 1735*, reprinted in Berkeley (1784))

Ward, John (1707) *The Young Mathematician Guide. Being a Plain and Easie Introduction to the Mathematics. In Five Parts. Viz. I. Arithmetic […] II. Algebra […] III. The Elements of geometry […] IV. Conick Sections […] V. The Arithmetic of Infinites* [...] *With an Appendix of Practical Gauging*, London: E. Midwinter, J. Taylor

Waring, Edward (1762) *Miscellanea Analytica, de Aequationibus algebraicis, et Curvarum Proprietatibus*, Cambridge: J. Bentham (1st chapter published in 1759)

Waring, Edward (1764) 'Problems', *Phil. Trans.*, LIII (1763), pp. 294–9

Waring, Edward (1765) 'Some new properties in conic sections', *Phil. Trans.*, LIV (1764), pp. 193–7

Waring, Edward (1766) 'Two theorems', *Phil. Trans.*, LV (1765), pp. 143–5

Waring, Edward (1770) *Meditationes Algebraicae*, Cambridge: J. Archdeacon

Waring, Edward (1772) *Proprietates Algebraicarum Curvarum*, Cambridge: J. Archdeacon (revision of (1762), part 2, chapters 1–4)

Waring, Edward (1776) *Meditationes Analyticae*, Cambridge: J. Archdeacon (2nd edn. 1785)

Waring, Edward (1779a) 'Problems concerning interpolations', *Phil. Trans.*, LXIX part 1 (1779), pp. 59–67

Waring, Edward (1779b) 'On the general resolution of algebraical equations', *Phil. Trans.*, LXIX part 1 (1779), pp. 86–104

Waring, Edward (1784) 'On the summation of series, whose general term is a determinate function of z the distance from the first term of the series', *Phil. Trans.*, LXXIV part 2 (1784), pp. 385–415

Waring, Edward (1786) 'On infinite series', *Phil. Trans.*, LXXVI part 1 (1786), pp. 81–117

Waring, Edward (1787) 'On finding the values of algebraical quantities by converging serieses, and demonstrating and extending propositions given by Pappus and others', *Phil. Trans.*, LXXVII part 1 (1787), pp. 71–83

Waring, Edward (1788) 'On centripetal forces', *Phil. Trans.*, LXXVIII part 1, (1788), pp. 67–102

Waring, Edward (1789a) 'On the method of correspondent values', *Phil. Trans.*, LXXIX part 2 (1789), pp. 166–84

Waring, Edward (1789b) 'On the resolution of attractive powers', *Phil. Trans.*, LXXIX part 2 (1789), pp. 185–98

Waring, Edward (1791) 'On infinite series', *Phil. Trans.*, LXXXI part 2 (1791), pp. 146–72

Waring, Edward (1799) 'Original letter of the late Dr. Waring to the Rev. Dr. Maskelyne, Astronomer-Royal', *The Monthly Magazine* VII, pp. 306–10

Waterland, Daniel (1730) *Advice to a Young Student, with a Method of Study for the Four First Years*, London: J. Crownfield

West, William (1761) *Mathematics. With Eleven Copper-Plates...*, London: by J. Kippax (2nd edn. 1763)

Whewell, William (1819) *An Elementary Treatise on Mechanics* [...] *vol* I *containing Statics and Part of Dynamics*, Cambridge: by J. Smith

Whewell, William (1823) *A Treatise on Dynamics. Containing a Considerable Collection of Mechanical Problems*, Cambridge: by J. Smith for J. Deighton

Whiston, William (1707) *Praelectiones Astronomicae Cantabrigiae in Scholis Publicis Habitae A Gulielmo Whiston, A. M. & Matheseos Professore Lucasiano. Quibus Accedunt Tabulae Plurimae Astronomicae Flamstedianae Correctae, Halleianae, Cassinianae, et Streetianae*, Cambridge: typis Academicis

Whiston, William (1710) *Praelectiones Physico-Mathematicae Cantabrigiae in Scholis Publicis Habitae. Quibus Philosophia Illustrissimi Newtoni Mathematica Explicatius traditur, & facilius demonstratur: Cometographia etiam Halleiana Commentariolo illustratur*, Cambridge: *typis* Academicis

Whiston, William (1716) *Sir Isaac Newton's Mathematick Philosophy More Easily Demonstrated: with Dr. Halley's Account of Comets Illustrated. Being Forty Lectures Read in the Publick Schools at Cambridge*, London: *for* J. Senex (English edn. of (1710) corrected and improved by the author)

Whiteside, Derek T. (1976) 'Newton's lunar theory: from high hope to disenchantment', *Vistas in Astronomy*, XIX, pp. 317–28

Wildbore, Charles (1791) 'On spherical motion', *Phil. Trans.*, LXXX part 2 (1790), pp. 496–559

Wood, James (1796) *The Principles of Mechanics: Designed for the Use of Students in the University*, in Vince and Wood, (1795–9), II

[Woodhouse, Robert] (1799) 'Théorie des Fonctions analytiques [...] by J. L. La Grange, of the National Institute. 4to. pp. 270. Paris', *The Monthly Review Enlarged*, XXVIII, pp. 481–99

[Woodhouse, Robert] (1800) 'Traité du calcul differentiel, &c, [...] by S. F. La Croix...', *The Monthly Review Enlarged*, XXXI, pp. 493–505; continued XXXII, pp. 485–95

Woodhouse, Robert (1801a) 'On the necessary truth of certain conclusions obtained by means of imaginary quantities', *Phil. Trans.*, XCI part 1 (1801), pp. 89–119

Woodhouse, Robert (1801b) 'Demonstration of a theorem by which such portions of the solidity of a sphere are assigned as admit of algebraical expression' *Phil. Trans.*, XCI part 1 (1801), pp. 153–8

Woodhouse, Robert (1802) 'On the independence of the analytical and geometrical methods of investigation; and on the advantages to be derived from their separation', *Phil. Trans.*, XCII part 1 (1802), pp. 85–125

Woodhouse, Robert (1803) *The Principles of Analytical Calculation*, Cambridge University Press

Woodhouse, Robert (1804) 'On the integration of certain differential expressions, with which problems in physical astronomy are connected', *Phil. Trans.*, XCIV part 2 (1804), pp. 219–78

Woodhouse, Robert (1809) *A Treatise on Plane and Spherical Trigonometry*, London: *sold by* Black, Perry and Kingsbury

Woodhouse, Robert (1810) *A Treatise on Isoperimetrical Problems and the Calculus of Variations*, Cambridge: *by* J. Smith, *sold by* Deighton, *et al.*

Woodhouse, Robert (1812) *An Elementary Treatise on Astronomy*, Cambridge: *by* J. Smith, printer to the university, *sold by* Black, Parry & co., Deighton

Woodhouse, Robert (1818) *An Elementary Treatise on Astronomy. vol. II Containing Physical Astronomy*, Cambridge: *by* J. Smith, printer to the University, *sold by* Black *et al.*

Woodhouse, Robert (1821) *A Treatise on Astronomy Theoretical and Practical, vol I*, Cambridge: *by* J. Smith, printer to the University, *for* Deighton and sons and G. W. & B. Whittaker

Wordsworth, Christopher (1877) *Scholae Academicae: Some Account of the Studies at the English Universities in the Eighteenth Century*, Cambridge University Press

Worster, Benjamin (1722) *A Compendious and Methodical Account of the Principles of Natural Philosophy: as they are Explain'd and Illustrated in the Course of Experiments, perform'd at the Academy in Little Tower-Street*, London: W. & J. Innys

Young, Matthew, (1784) *An Inquiry into the Principal Phaenomena of Sounds and Musical Strings*, Dublin: by J. Hill

Young, Matthew (1800) *An Analysis of the Principles of Natural Philosophy*, Dublin University Press, *by and for* R.E. Mercier and G.G. and J. Robinson

INDEX

Simpson, Thomas, ix, x, xi, xii, 56ff, 63, 66, 68, 69, 73ff, 79, 82ff, 89, 101, 109, 110, 115, 125, 127, 131, 138, 139, 148, 156, 160, 161, 162, 163, 164, 175n, 176n, 177n, 178n, 180n
Simson, Robert, xi, 12, 25, 26, 27, 36, 37, 95, 96, 99, 101, 149, 153, 161, 166, 169n, 170n, 175n, 177n, 179n
Sinclair, George, 152
Sinclair, Robert, 152-3
Sloane, Hans, 14
Sluse, René François de, 13
Smeaton, John, 160
Smith, G. C., xi, 46
Smith, James, 46, 161, 173n
Smith, John, 152, 170n
Smith, Robert, 23, 27, 29, 30, 31, 85, 125, 151, 163, 170n, 171n
Smyth, John, 179n, 180n
Spalding Gentleman's Society, 62, 65
Spence, William, 95, 104ff, 120, 135, 136, 137, 140, 162, 179n, 182n
Spitalfields Mathematical Society, 65
Stack, J., 132, 163
Stewart, Dougald, 98, 99, 152, 175n
Stewart, John, 57, 62, 66, 96, 154, 166, 170n
Stewart, Matthew, ix, xi, 25, 26, 37, 98, 101, 152, 160, 161, 175n, 176-7n
Stirling, James, ix, x, xi, xii, 28, 33ff, 42, 43, 48, 69, 83, 139, 160, 161, 163, 168n, 169n, 171n, 172n, 173n
Stone, Edmund, xii, 17, 41, 55, 58, 61, 161, 162, 164, 167n, 169n, 176n
Swall, A., 179n

Taylor, Brook, ix, x, xi, xii, 21, 28, 31ff, 42, 43, 83, 139, 159, 160, 162, 168n, 169n, 171n
Taylor, E. G. R., xi
Taylor, W. B. S., 181n
teachers of mathematics, xi, 14, 16, 27, 33, 61-2, 65, 111
Thomas, Hugh S., 180n
Thomson, Thomas, 119
Thorp, Robert, 166
Todhunter, Isaac, 176n, 177n, 178n
Toplis, John, 116, 118, 135, 161, 179n
Townshend, George, 179n
Trial, William, xi, 25, 96-7, 99, 154, 159, 169n, 179n
translations, x, 17, 57, 78, 110, 118, 167n, 169n
Tripos Exam, 25, 63, 116, 126, 181n
Truesdell, Clifford A., 176n
Tschirmhaus, E. W., 13, 169n

Turnbull, Herbert W., xi, 173n, 174n
Turner, G. L' E., 21, 170n
Tweedie, Charles, xi, 34, 35, 171n, 172n, 173n, 174n

universities: Aberdeen, 96-7, 153-4, 170n; Cambridge, 12, 22ff, 116, 122, 124ff, 141, 150-1, 181n; Dublin, 132ff, 141, 155; Edinburgh, 12, 26ff, 98, 152; Glasgow, 12, 25, 95-6, 152-3; Oxford, 12, 19ff, 63, 151-2; St Andrews, 97, 153, 170n
Ussher, Henry, 132, 155

Varignon, Pierre, 21, 168n, 169n
Vauban, Sébastien Le Prestre de, 110, 114, 179n, 180n
Venturoli, Giuseppe, 134, 162
Vilant, Nicolas, 97, 153
Vince, Samuel, 56, 62, 64, 124ff, 151, 160, 161, 162, 163, 164, 175n, 181n
Vream, W., 171n

Waddington, Robert, 158
Waff, Craig B., 176n, 177n
Wales, William, 169n
Walker, Adam, 181n
Wallace, John, 157
Wallace, William, xiii, 95, 98, 103-4, 108, 114, 116, 117, 120, 126, 133, 135, 139, 157, 161, 164, 179n, 180n, 181n
Wallis, John, ix, 11, 12, 13, 19, 23, 151, 167n, 168n, 169n
Wallis, Peter J., xii, 64, 66, 171n, 175n, 176n, 181n
Wallis, Ruth V., xii, 66, 175n, 176n, 181n
Walmesley, Charles, 133, 160, 161, 176n
Walter, Thomas, 162, 176n
Walton, Jakob, 46, 131, 161, 173n
Ward, John, 160, 164
Waring, Edward, ix, x, xii, 27, 64, 82, 85, 89ff, 101, 124, 126, 127, 132, 138, 139, 151, 159, 160, 161, 162, 163, 179n
Waterland, Daniel, 22-3, 170n
Watson, Charles, 179n
Watson, Henry, 113
Watt's Academy, 33, 171n
Wentworth, William, 178n
West, John, 97, 103
West, William, 160
Whewell, William, 134, 162
Whiston, William, 18, 20, 21, 22, 23, 24, 65, 150, 160, 163, 170n, 176n

Whiteside, John, 21
Whiteside, Tom, xi, xiv, 11, 169n, 177n
Wildbore, Charles, 162
Williams (Colonel), 180n
Williams, Thomas, 176n
Williamson, James, 25, 95–6, 153
Williamson, Joseph (Free Mathematical
  School), 27, 62
Wilson, James, 24, 26, 175n
Witchell, George, 123, 158

Wood, James, 64, 125, 162
Woodhouse, Robert, xiii, 64, 85, 105,
  117, 120–1, 126ff, 132, 133, 135, 136,
  138, 139, 151, 160, 161, 162, 164,
  179n, 181n, 182n
Wordsworth, Christopher, 63, 64, 170n
Worster, Benjamin, 24, 33, 163, 171n

Young, John, 152, 159, 163
Young, Matthew, 132